D0929101

The Kentucky Bicentennial Bookshelf
Sponsored by
KENTUCKY HISTORICAL EVENTS CELEBRATION COMMISSION
KENTUCKY FEDERATION OF WOMEN'S CLUBS
and Contributing Sponsors

AMERICAN FEDERAL SAVINGS & LOAN ASSOCIATION
ARMCO STEEL CORPORATION, ASHLAND WORKS
A. ARNOLD & SON TRANSFER & STORAGE CO., INC. / ASHLAND OIL, INC.
BAILEY MINING COMPANY, BYPRO, KENTUCKY
BEGLEY DRUG COMPANY / J. WINSTON COLEMAN, JR.
CONVENIENT INDUSTRIES OF AMERICA, INC.
IN MEMORY OF MR. AND MRS. J. SHERMAN COOPER BY THEIR CHILDREN
CORNING GLASS WORKS FOUNDATION / MRS. CLORA CORRELL
THE COURIER-JOURNAL AND THE LOUISVILLE TIMES
COVINGTON TRUST & BANKING COMPANY
MR. AND MRS. GEORGE P. CROUNSE / GEORGE E. EVANS, JR.
FARMERS BANK & CAPITAL TRUST COMPANY
FISHER-PRICE TOYS, MURRAY
MARY PAULINE FOX, M.D., IN HONOR OF CHLOE GIFFORD
MARY A. HALL, M.D., IN HONOR OF PAT LEE,
JANICE HALL & MARY ANN FAULKNER
OSCAR HORNSBY INC. / OFFICE PRODUCTS DIVISION IBM CORPORATION
JERRY'S RESTAURANTS / ROBERT B. JEWELL
LEE S. JONES / KENTUCKIANA GIRL SCOUT COUNCIL
KENTUCKY BANKERS ASSOCIATION
KENTUCKY COAL ASSOCIATION, INC.
THE KENTUCKY JOCKEY CLUB, INC. / THE LEXINGTON WOMAN'S CLUB
LINCOLN INCOME LIFE INSURANCE COMPANY
LORILLARD A DIVISION OF LOEW'S THEATRES, INC.
METROPOLITAN WOMAN'S CLUB OF LEXINGTON
BETTY HAGGIN MOLLOY
MUTUAL FEDERAL SAVINGS & LOAN ASSOCIATION
NATIONAL INDUSTRIES, INC. / RAND MCNALLY & COMPANY
PHILIP MORRIS, INCORPORATED / MRS. VICTOR SAMS
SHELL OIL COMPANY, LOUISVILLE
SOUTH CENTRAL BELL TELEPHONE COMPANY
SOUTHERN BELLE DAIRY CO. INC.
STANDARD OIL COMPANY (KENTUCKY)
STANDARD PRINTING CO., H. M. KESSLER, PRESIDENT
STATE BANK & TRUST COMPANY, RICHMOND
THOMAS INDUSTRIES INC. / TIP TOP COAL CO., INC.
MARY L. WISS, M.D. / YOUNGER WOMAN'S CLUB OF ST. MATTHEWS

The Kentucky Harness Horse

KEN McCARR

Foreword by Larry Evans

THE UNIVERSITY PRESS OF KENTUCKY

McCarr, Ken, 1903-1977.
The Kentucky harness horse.

(The Kentucky Bicentennial bookshelf)
Bibliography: p.
1. Harness racehorses—Kentucky—History.
2. Harness racing—Kentucky—History. I. Title.
II. Series.
SF339.5.U6M3 1978 636.1′3 75-3548
ISBN 0-8131-0213-8

Research for The Kentucky Bicentennial Bookshelf
is assisted by a grant from the
National Endowment for the Humanities.
Views expressed in the Bookshelf do not
necessarily represent those of the Endowment.

Scholarly publisher for the Commonwealth,
serving Berea College, Centre College of Kentucky,
Eastern Kentucky University, The Filson Club,
Georgetown College, Kentucky Historical Society,
Kentucky State University, Morehead State University,
Murray State University, Northern Kentucky University,
Transylvania University, University of Kentucky,
University of Louisville, and Western Kentucky University.

Editorial and Sales Offices: Lexington, Kentucky 40506

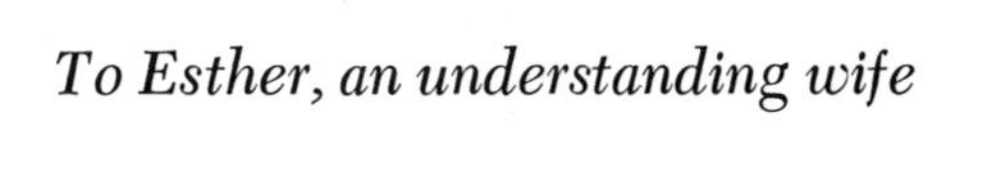
To Esther, an understanding wife

Contents

Foreword ix

Acknowledgments xi

A Prefatory Note xii

1 / Beginnings 1

2 / Systematic Breeding and Early Speed 15

3 / The George Wilkes Boom 25

4 / Spread of the Kentucky Influence 34

5 / Four Families of Trotting 47

6 / Patriarchs of Pacing 69

7 / Belgravian Dames 81

8 / Producers of Champions 99

9 / Kentucky's Trotting Races 115

A Note on Sources 131

Color illustrations follow page 68

Foreword

THE PUBLISHERS of this series could have found no finer historian to chronicle the harness horse's past in Kentucky than my friend Ken McCarr, who died in May, 1977, after a lifetime in the harness racing sport.

Ken's more than seven decades were spent almost entirely in harness racing. Several of his boyhood years were spent at the Savage Farm in Minnesota where his father, Ned, handled the colts while stable star Dan Patch was barnstorming around the country. To the end of his life Ken cherished affectionate memories of Dan Patch, and sometimes dreamed of writing a book about him someday.

When Ned McCarr moved to Pennsylvania to train for J. D. Callery of Pittsburgh and was headquartered at Arden Downs near Washington, Pennsylvania, young Ken graduated from high school there and attended Washington and Jefferson College for two years. From there he went to Central State Teachers College, now the University of Wisconsin at Stevens Point. While there, in 1926, he did his first turf writing when he covered winter training activities for *Trotter & Pacer* magazine, and during a break in his college career he was a full-time writer.

After graduation, McCarr taught high school and coached basketball and track in two Wisconsin towns and served as principal at Lublin. Ailing, perhaps from the confinement, McCarr then returned to the horses as a groom and worked with several leading harness racing stables including those trained by Bill "Chilly Willy" Hodson and Walter Cox and the famous Arden Homestead stable headed by Harry Pownall.

After he left teaching and returned to the track, McCarr continued as correspondent for turf weeklies. While at Pinehurst, North Carolina, for winter training in 1934, he met a nurse who had never seen a horse race, and as he explained the sport to her a mutual admiration developed. Esther McCarr was part of the team for more than forty years. Pony-sized in comparison to Ken's rugged build, Esther still could keep chaos from setting in when Ken was scattering record books seeking the great-granddam of the sire of a current trotting champion.

After World War II he joined the staff of the *Horseman & Fair World* magazine, then published in Indianapolis. He rose rapidly and became editor in 1948, only to join the United States Trotting Association shortly thereafter as registrar, a position he held for more than twenty-two years.

When he retired on January 1, 1971, Ken looked forward to a time of leisure. As so often happens, he found himself as busy as ever writing for three magazines, *Hoof Beats, The Harness Horse,* and *Hub Rail,* with special assignments in Australia, New Zealand, Italy, and Sicily. He had probably the largest private library of harness horse books and magazines in the world, and McCarr's historical research was both accurate and comprehensive. One of his last major projects, and one in which he took an especially keen interest, was this volume for the Kentucky Bicentennial Bookshelf.

LARRY EVANS

Acknowledgments

IN HIS MANUSCRIPT, the late Ken McCarr acknowledged photographs from the United States Trotting Association, *The Harness Horse, The Horseman & Fair World, The Thoroughbred Record*, and George Smallsreed, Jr. Some of the photographs were copied for him by Jan Brown.

The publisher thanks Tom White for his very considerable help in revising and preparing the manuscript for the press after the author's death. We are also grateful for the expert advice of Thomas R. Culbertson, and to the staffs of *The Horseman & Fair World* and The Red Mile. Dr. Mary Wharton shared with us her extensive knowledge of Bluegrass farm history.

The old prints that appear in the color section were copied for us by Phil Pines of the Hall of Fame of the Trotter in Goshen, New York. The oil painting of the first grandstand at the Lexington Trotting Track is by Mary C. Nickell and is the property of The Red Mile. It was photographed by Clyde Burke, who also provided the Bluegrass pasture scene, the picture of the Floral Hall, and the Red Mile racing scene. The clubhouse scene is by George Smallsreed, Jr.

A Prefatory Note

THIS BOOK is addressed especially, with a note of welcome, to those who are newly acquainted with the traditional American sport of harness racing and who may wish to acquire some knowledge of its history.

So far as possible, therefore, the author has cleared away the forest of names and figures so dear to harness racing veterans.

1

BEGINNINGS

The long afternoon shadows of October, 1974, were stealing across the dusty old clay surface of the Lexington trotting track. The greatest group of two-year-old pacers ever assembled was battling out an important race. Among them were Nero, undefeated in twelve races, and Alert Bret, who already had eight two-minute miles to his credit. Nero led to the quarter, yielded for a moment to Farthebest Hanover, then reached the half in :56 3/5. Then Alert Bret moved up to challenge him, matching Nero stride for stride until he led by a neck at the head of the Red Mile's long homestretch. Nero fought his way back to even terms in mid-stretch, but Alert Bret came on again and held on to win by a neck. The time of the mile was a world record–1:55 4/5. No one who saw it will ever forget that race.

This was typically Lexington. For a century the old track has been the scene of fast contests among great young horses, many of them born in Kentucky. Just a year earlier these same youngsters had nibbled on the bluegrass and frolicked in the pastures of nearby farms. Now they were setting world records; they had come a long way from the time when the pioneers first brought horses through the Cumberland Gap.

Today's harness horses race at two gaits–the trot and the pace. On the trot, the left front and right rear legs move forward in unison, then the right front and left rear; so trotting is called the diagonal gait. A pacer moves both left legs forward together, then both right legs. Both of these gaits are natural ways of going. But attaining high speed at either gait, and the ability to maintain the gait under the pressure of competition without breaking into a run, require special talents (fostered by selective breeding) and skilful training. Americans began to be interested in these qualities about the beginning of the nineteenth century and their interest culminated in the creation of a distinctively American breed of racehorse, now called the Standardbred. The Standardbred horse is a derivative of the Thoroughbred; Thoroughbred forebears are prominent in all Standardbred pedigrees. But, as later chapters will show, other elements are present also.

Racing arose from the competitive urge in man. An attempt of one rider to pass another meant a contest over the primitive roads. Later—when carriage roads had been built—young lovers out for a buggy ride would do a little hot-rodding when another vehicle approached, to impress their girl friends. Often, though, the horse's only competitor was a stopwatch; from the first, *time* seems to have been basic to the development of the Standardbred. (And ultimately, time over a mile distance was the standard for registration.) A wager would be made, for instance, that a horse could trot over a measured mile distance in three minutes. Later on, it was by the standard of speed—measured in minutes, seconds, and fractions for a mile distance—that the ancestors of today's harness racehorses were selected.

Some early contests were match races, in which an owner matched the speed of his horse against a rival for an agreed amount of money. Some of these impromptu contests were held on the streets of the village of Goshen, New York, as early as Revolutionary War times, but the results of these early races have been lost. Sometimes the early trotters raced under saddle; then came the light four-wheeled wagons, and in turn the high-wheeled sulky, a two-wheeled cart that weighed ninety pounds or more. The driver sat high above and immediately behind the horse in this cart. The big wheels were made of ash wood. Sometimes, in collisions, they splintered, but there is no record of one of these big wheels collapsing.

An old high-wheeled sulky compared with a modern, lighter vehicle

When the bicycle craze hit the nation in 1892, a new light cart with pneumatic tires appeared. This made the cumbersome high-wheeled sulky obsolete, for the new sulky weighed only about a third as much. This light sulky gave a boost to the steadily increasing speed of the harness horses, which still doesn't seem to have peaked.

Until 1792, Kentucky was the western part of Virginia, which was the first of the original thirteen colonies to import horses. These first horses arrived there in 1609, just two years after the landing of the settlers at Jamestown. A famine that winter led to the demise of both the horses and men, "they having fed on horses and other beasties as long as they lasted," as one of the pioneers wrote. Later other horses were brought to the New World for labor and transportation. A visiting Englishman described these horses as pacers—small, hardy, strong, and fleet. The pace is a comfortable gait for a rider. Nearly everyone in the colonies had a horse and the animals were turned loose to roam at will.

Even though horses in the colonies were chiefly used for farm work or transportation, racing began early. In 1677 there were three trials in the Henrico County, Virginia, court concerning people who had not honored their wagers on races. George Washington, who kept very exact records, noted that he paid Robert Sanford 12 shillings, about $3, to ride his pacer in a race at Accotink, near Mount Vernon, on August 29, 1768; no winnings on this race are mentioned. Washington evidently had as much trouble picking a winner as the rest of us. After the races at Annapolis on October 10, 1772, his ledger showed, "Cash lost on the races 1.6 pounds."

The Revolutionary War halted the importation of horses from England, at the same time slowing the rise of organized racing. But when the war was over the traffic in horses began again. One of the most important Thoroughbreds to be brought to the United States at this period was Messenger, a gray, who had started racing in England in 1783 at the age of three and in the ensuing three years won eight of the fourteen races in which he started. Thomas Benger, sometimes called Sir Thomas, an Irish sportsman and country squire, emigrated to America in 1788 and brought Messenger with him. Two grooms were needed to control the stallion as he charged down the gangplank at Philadelphia. Three years later, a severe yellow fever epidemic ravaged the Philadelphia area and Benger hurriedly disposed of all of his holdings—including Messenger—and fled back to Ireland. Under several ownerships, Messenger was then moved over eastern Pennsylvania, New Jersey, the lower part of New York State, and Long Island, where he died in 1808. He had little opposition from other stallions and was the most popular horse in the North.

No attempt was ever made to cultivate a trot in Messenger, but at times he would break into an impressive trot while being shown. Several of his sons were the best of their day in Thoroughbred circles, and some of these were also to exert a great influence on Kentucky's harness horses. Today Messenger is an ancestor of many of the best American Thoroughbreds and all Standardbred horses trace to him on the paternal side of the pedigree. His sons and daughters founded distinct families that became the greatest in both breeds.

Meanwhile, the new nation was expanding westward. The hardy pioneers who went through the Cumberland Gap brought horses with them, but they had no thought of racing, for they went into a wilderness called the Dark and Bloody Ground. To them the horse was a means of transportation, and in the trackless wilderness the best were those that were sturdy and comfortable under the saddle. The tough little pacers of the coastal colonies, and perhaps some Thoroughbreds, were the companions of the first settlers.

Kentucky frontiersmen, after seeing cattle and sheep fatten on the lush wild grass for which the Bluegrass area was named, discovered that it was equally good for horses. The limestone underlying this area caused the ample, pure water to be filled with minerals beneficial to the building and strengthening of the light bones of fast horses, and Kentucky's long summers and mild winters were also ideal for raising horses. Thoroughbred men were the first to recognize these advantages and their horses overshadowed the trotters and pacers for many years. Today there are still some who associate only Thoroughbreds with Kentucky, but Kentucky's harness horses also were to find a place of worldwide prominence.

Until 1850 there were no trotting races south of the Pennsylvania line except in New Orleans. Thus in the first half of the nineteenth century there was no motivation to cultivate the trotting horse in Kentucky. When thought began to be given to the trotters, it was not racing that accounted for the change. As the population increased and grew more prosperous, roads were built and a demand for fine carriage horses arose. John Wesley Hunt of Lexington (who was to become Kentucky's first millionaire) was the first to recognize the opportunity. He knew that some lines from imported Messenger were producing fine trotters in New York, and he thought this would be the ideal bloodline to bring into Kentucky to breed light harness horses for use on the new pikes.

Hunt got in touch with William T. Porter, editor of the first weekly journal devoted to sports, the *Spirit of the Times.* This journalist was supposed to be an absolute authority on horses. Hunt authorized him to select two of the best stallions of the Messenger line for him. Porter chose Abdallah and Commodore, both by a son of Messenger, Mambrino. The sons of Mambrino were getting scarce at that time because of exports by the British for improving the cavalry in the West Indies, and also because they were sought by buyers from the surrounding states for the production of carriage horses. So neither Abdallah nor Commodore came cheap.

In February of 1840, in the worst part of winter, Abdallah and Commodore arrived in Lexington after being ridden overland. Commodore was a big, showy horse that attracted attention. The breeders accepted him, but he never sired any trotter of note in his Kentucky home. Abdallah, who had cost $1,000, was not an attractive horse and in addition to his ugliness had a vile disposition. He was not received with favor by Kentucky breeders. Although he had been brought in to improve Kentucky driving horses, Abdallah himself could not be driven. A Lexington trainer, Dennis Seals, tried to subdue Abdallah by placing a harness on him and turning him out in a paddock where he was allowed to kick to his heart's content. In the end Abdallah was the victor—he was never driven.

Abdallah's ugliness, both in looks and deportment, coupled with the fact that he was allowed to serve only a few mares, made Hunt anxious to get rid of him. He expected to take a heavy loss, but it worked out exactly the opposite, for William Simonson, a New York City butcher, was willing to pay $1,365 for the stallion. The early foals of Abdallah had been showing speed and the New Yorkers wanted him back.

The rejection of Abdallah by the Kentuckians may well have robbed the state of the title of "cradle of the trotter," for Abdallah went back to New York and, at the age of twenty-six, sired Hambletonian, who is called "the great father of trotters." Hambletonian and his sons came from Orange County, New York. Today there is no Standardbred horse, either trotter or pacer, that does not trace back to Hambletonian several times.

The first successful trotting sire in the Bluegrass country was a grandson of Mambrino, and a son of Henry Clay was responsible for bringing that horse, Mambrino Chief, to Ashland. James B. Clay had no desire to follow in his father's footsteps politically, but he did share Henry Clay's affinity for Thoroughbreds and his fancy for breeding fine cattle. Many good horses had been quartered in the Ashland stables during Henry Clay's lifetime. When the great statesman died in 1852, James paid $63,760 for the Ashland mansion and 337 acres of the original 600-acre tract. It was his interest in cattle that led James Clay to travel to Dutchess County, New York, in 1853. Among Clay's many friends was Edwin Thorne, owner of the Thorndale Farm at Millbrook, near Poughkeepsie. Thorne had the most famous shorthorn herd in America and Clay had hopes of obtaining new blood there as an outcross for his cattle at Ashland. But at Thorndale he saw something new to him—trotters of the Messenger line, which were being bred in that region not just for carriage horses, but for racing.

The only known sketch of MAMBRINO CHIEF, made after his death

Clay now made a decision that was to create a furor in the Bluegrass. The Virginians who had settled Kentucky not only had brought their horses with them but also their prejudices. They believed that the only horse worthy of attention as a racehorse was the Thoroughbred, and the aristocracy of the Bluegrass did everything possible to hinder the introduction of the sport of harness racing to Kentucky. When James Clay decided to enter into the production of trotters for racing purposes this group was placed at a distinct disadvantage. Against the prestige and wealth of the Clay family no open fight could be waged, but for many years there was bitter opposition to James Clay's enterprise.

An element of luck here enters the story, for Mambrino Chief—who was to be the progenitor of many fine Bluegrass horses—was not Clay's original choice to become Ashland's foundation sire. He had instructed Thorne to purchase Burr's Washington for $3,000. Thorne, however, recommended Mambrino Chief at $4,000 as a better choice. Clay agreed, even though the price was rather steep for an untried sire and Clay had heavy financial burdens at that time. (The Ashland mansion was in poor repair and had to be completely rebuilt.) Josiah Downing was engaged to bring the stallion to Kentucky. After being ridden all the way from Dutchess County, Mambrino Chief arrived at Ashland on February 21, 1854.

The appearance of a new stallion is likely to create rivalries. At that time there was one noted trotting sire in the Bluegrass, called Pilot Jr. He could hardly be called trotting bred, but he had quite a following nevertheless, and a challenge was brought forward for a match race with Mambrino Chief, the horses to go two-mile heats for a bet of $1,000 between the owners. Although the Chief had just completed a heavy stud season of 80 mares, he was placed in the charge of a famous trainer, Dr. Levi Herr. This Pennsylvanian had become a permanent resident of Lexington at the age of thirty-eight. His skill in veterinary medicine, added to his great abilities as a trainer and breeder, had made Dr. Herr one of the most honored citizens of Lexington.

In spite of his bad feet, Mambrino Chief could show such bursts of speed that the Pilot Jr. camp realized that their entry would suffer by comparison, so they paid the forfeit. Both of these early Kentucky stallions were later to become noted for their great daughters.

No portraits were made of Mambrino Chief during his life, for his real reputation came after his death on March 28, 1862, when he was eighteen years old and standing in Woodford County. An artist by the name of Schultz later made a sketch, relying on the descriptions given by persons who had known Mambrino Chief. This purported likeness, published in Chester's *Complete Trotting and Pacing Record*, is the only known picture of Kentucky's first great trotting sire.

It was Mambrino Chief that started Kentucky on its way to becoming a great producer of trotting horses. At one time the owners of trotters sired by Mambrino Chief, along with the many other supporters of the stallion, vehemently maintained that Mambrino Chief was superior to Hambletonian.

History is loaded with attempts at deflation and those by horsemen were as bad as those found in politics. One of the favorite allegations was that a famous horse had a different sire from the one credited to him. Mambrino Chief fell afoul of one of these hate campaigns forty years after his death. The first gun was fired by General John B. Castleman, for many years the head of the Kentucky Saddle Horse Association. He said that Mambrino Chief was "coarse headed, coarse eared, coarse legged and coarse hoofed and no fine colt was ever sired by Mambrino Chief."

But even after forty years had elapsed, there was still someone with authority who had known the horse. The allegation brought a rebuttal from the son of James B. Clay. His description was radically different and wound up with the sentence, "He was a large horse but I never thought him a coarse one."

The detractors also singled out Lady Thorn, a daughter of Mambrino Chief who was one of the greatest racing mares of her day. (She appeared to have the world trotting record at her mercy when her racing career was suddenly terminated by an accident. The portable loading ramp slipped while she was being loaded on a railway car and she fell heavily, hitting a rail and breaking her hip.)

It was alleged that Lady Thorn was not by Mambrino Chief but by the saddle stallion Gaines' Denmark, which had been standing at the same location where the Chief was doing stud duty. The story was that the dam of Lady Thorn was secretly mated with the saddle stallion. The breeder, Dr. Herr, vigorously denied this, but it was not possible at that time to prove conclusively that his mare was by Mambrino Chief. Later research, however, proved that Herr was correct.

An unknown mare, "which had come from the west" in a drove of horses, was the dam of Mambrino Chief. The drove came down a valley from Pennsylvania to New York. In the Pennsylvania neighborhood from which the mare had come, there was a stallion called Messenger Duroc and those well versed in the strain from this stallion saw a resemblance in the mare. Messenger Duroc passed on two oddities to his offspring. First, those of a certain color were apt to have large feet that were always tender. The dam of the Chief had big feet and he himself could not stand heavy work, for his feet would crack open and get sore, like a fingernail broken to the quick. This weakness he also passed along to his foals. The second oddity was that foals of Messenger Duroc had a gray tone on their legs. The right hind leg of Mambrino Chief was a steel gray. The Chief also passed this along to his foals and it bobbed up in later generations as well. The point overlooked by the detractors was that Mambrino Chief, like the old silversmiths, had put a hallmark on his product: there were white hairs on the right hind leg of Lady Thorn. The first racing trotter from Kentucky to attain fame was indeed sired by Mambrino Chief.

The best-known son of the Chief was Mambrino Patchen, one of the last foals to be sired by the old stallion. Patchen was foaled in the stable lot at Dr. Herr's Forest Park Farm in 1862, after Mambrino Chief's death. The stallion was a full brother to Lady Thorn and, unlike many of the offspring of Mambrino Chief, he was a handsome individual. His good looks again raised the claim that he was by a saddle stallion but, like his sister, he carried the mark of the Chief, for his only markings were the white hairs on his hind legs.

Dr. Herr was standing Mambrino Pilot, another son of Mambrino Chief, when Mambrino Patchen was born. The following year Patchen was sold for $1,500, a price that no yearling had ever before commanded. Then Dr. Herr sold Mambrino Pilot to eastern interests for a fancy price, never made known. Now without a stallion, he repurchased Mambrino Patchen and returned him to Lexington.

As a three-year-old, Mambrino Patchen was given his first lessons in harness and was bred to a few mares. The resulting foals arrived during the Civil War and all traces of them were lost. Patchen was a magnificent stallion but he was never shown in harness, so rival stallion owners were quick to claim that the horse had no speed. Regardless of this, Mambrino Patchen did sire foals that attained speed at an early age. One, Lady Stout, was the first three-year-old to trot a mile in less than 2:30, a championship performance in 1874.

The foals of Mambrino Patchen were trotters and Dr. Herr stipulated that he would refund the service fee on any foal that preferred the pacing gait. He only had to make two or three refunds. This horse not only became the first stallion to sire this early speed but he was also noted as a broodmare sire, for his daughters were the dams of some of the fastest horses in their generation. His best-known daughter was Alma Mater, the greatest broodmare of her time. Her sons were champions on the track and later, as sires, they helped to upgrade the trotting breed.

One of Alma's sons sired a horse registered in the American Trotting Register as Allan, a foal of 1886, who played a prominent part in the formation of the breed of Tennessee Walking Horses. As a five-year-old he was sold for $350 in a mixed sale at Brasfield & Co. and was taken to Murfreesboro, Tennessee, where he temporarily disappeared from sight. But later a man named James Brantley, a Tennessee breeder, was intrigued by a good-looking horse that had been traded for an $80 calf. He rode long distances and did a great deal of searching to find the pedigree of this obscure horse, and eventually he even found the registration certificate. He then bought Allan for $110. This horse is now the number one foundation sire of the Tennessee Walking Horse breed and head of the "Allen" strain. No one is sure how the name came to be altered, but it was probably due to a misunderstanding by one of the stallion's many owners.

Mambrino King, another son of Mambrino Patchen, was a noted show horse and stood for service at Dr. Herr's farm. Overshadowed by his sire, he had little access to the good trotting matrons although his handsomeness drew saddle and show mares. But Mambrino King attracted the notice of C. J. Hamlin, who had made a fortune in sugar and then established the famous old Village Farm at East Aurora, New York. In the spring of 1882 Hamlin weeded out his breeding stock and headed for Kentucky to get a better class of horses. Hamlin demanded beauty in a horse and had commented, "When you go into a ballroom you would much rather choose as a partner a beautiful woman that can dance well than a homely one that can dance equally well." He paid $17,000 for Mambrino King and labeled his new purchase "the handsomest horse in the world." Mambrino King did have an outstanding record in the show ring and was only beaten twice, and then by Kentucky-bred horses. Mambrino King's most lasting fame came through a daughter, Nettie King, who produced the champion trotter The Abbot and his full brother, The Abbe, who preferred the pacing gait. Today The Abbe is prominent for his part in starting one of the greatest male lines of pacers, for from this came the immortal Adios, the first harness horse to sell for half a million dollars. The number of Adios's fast sons and daughters was unequaled for many years, but in 1974 was matched by a sire called Tar Heel, whose opportunities were enhanced by being crossed on the daughters of Adios.*

---

* In May, 1977, Meadow Skipper, which stands at Stoner Creek Stud in Bourbon County, Kentucky, surpassed both Adios and

The line of Mambrino Chief had left an indelible mark on two breeds of horses but it was not through yet. One son, Clark Chief, took his name from his place of birth, Clark County, Kentucky. Few of the sons and daughters of this horse were used for breeding purposes but from this small group came Harrison Chief. Although Harrison Chief may never have had a saddle on him in his life, for he was strictly a trotter, he created the Chief family of saddle horses. When the Saddle Horse Breeders Association was formed he was included on a list of the stallions that had created the foundation of that breed. In 1901 he was removed from the list of foundation sires, for this was at the time when General John B. Castleman tried to ignore any family that traced back to the horse he so hated, Mambrino Chief. In spite of this demotion, Harrison Chief is still recognized as the founder of the Chief family of saddlers.

Many Kentucky horses owe a debt of gratitude to James B. Clay for bringing Mambrino Chief to the Bluegrass.

Tar Heel as the leading sire of two-minute performers and became the first Standardbred stallion to sire 100 two-minute sons and daughters. Meadow Skipper is also a descendant of The Abbe. Ed.

2

SYSTEMATIC BREEDING AND EARLY SPEED

In frontier days, the breeding of horses was a hit-or-miss proposition; the owner of a mare would take her to the nearest stallion or the horse with the cheapest service fee. In the spring stallion owners would move their horses from village to village and set up headquarters at the local livery stable. Competition was keen and often the mare owners were influenced by the jug of homemade liquor, and/or a reduction in the established service fee.

Systematic breeding was first introduced at Woodburn Farm, the greatest of the early Kentucky breeding establishments. It was the first farm in the nation to match the greatest mares with the greatest stallions and plan the crossing of bloodlines to produce the greatest speed in the resulting foals. The principles introduced at Woodburn still govern the great farms of both the Thoroughbreds and the Standardbreds.

Woodburn was part of a large United States land grant to General Hugh Mercer for services in the Revolutionary War. Mercer's tract of 3,200 acres was located about ten miles east of the present city of Frankfort, at Spring Station. The land that became Woodburn was purchased by Robert Alexander, a member of a wealthy Scottish family. While a student in France, Alexander happened to strike up an acquaintance with Benjamin Franklin, then the United States ambassador to France. Franklin took a liking to the young Scotsman and hired him as his secretary. When Franklin returned to America, he was followed shortly afterwards by Alexander. Their warm friendship continued until Franklin's death in 1790 and then Alexander decided to head west. Alexander purchased Woodburn and located there shortly before Kentucky was admitted to the Union in 1792. He was prominent in local affairs and became the state senator from Woodford County and also president of the Bank of Kentucky, located at Frankfort. This brought about a need for a second home at Frankfort, where he met his wife, Eliza Weisinger. Their son, Robert Aitcheson Alexander, who was to become famous as a horseman, was born at Woodburn in 1819. He was sent to England for his education, and there young Robert became interested in the breeding of cattle, sheep, and other farm animals. He studied correct soil treatment, drainage, croppage, and the most approved methods of raising and housing livestock. At this time he had no idea of racehorse production, but he returned to his native Kentucky a better-equipped gentleman farmer than any other Kentuckian up to that time.

When Woodburn passed to R. A. Alexander upon the death of his father, the young man put his extensive knowledge to work in building up the farm. He became a permanent resident of Woodburn in 1851 and for five years raised mostly cattle and sheep. Then his interest turned to horses. With the advantage of ample funds, it took only a few years for him to attain the top position among American breeders and to make Woodburn Farm the foremost breeding establishment in the nation for both runners and harness horses.

The great stallion Lexington had just completed his racing career and Alexander selected him to head his Thoroughbred department. The $15,000 he paid for Lexington was the highest that had ever been paid for an American horse, regardless of breed. The enthusiasm of the Thoroughbred people for this new venture at Woodburn was somewhat dampened when it was learned that Alexander was following James Clay in the production of trotters. Here again, Alexander's wealth enabled him to collect the best breeding stock. One of his first selections was the broodmare Madame Temple, dam of the champion trotter of the day, Flora Temple; Flora was the "bob-tailed mare" of Stephen Foster's song about the Camptown races.

The first year of Woodburn's venture into the horse business was 1856 and, although the trotters were not shown in his first farm catalog, that was the year that Alexander purchased his first trotting sire, Edwin Forrest. One year later he bought a pair of colts by Mambrino Chief—Bay Chief and Brown Chief—and his first famous stallion, Pilot Jr.

It is a mystery why Pilot Jr. was selected to stand at Woodburn, for the horse could not be called fashionably bred. His dam, Nancy Pope, had a shady background; several pedigrees had been given for her, naming as her sire everything from a farm horse to the Thoroughbred stallion Havoc, who had died four years before Nancy was born. No smart horseman would have picked a son of this mare to be a great sire. But Pilot Jr.'s sire was a Canadian pacing stallion called Pilot, Pacing Pilot, or occasionally Old Pilot, and he was another story.

Pilot was a rugged black horse bought for $125 in Montreal by an itinerant peddler named Elias Rockwell. The horse had great speed but at times it was almost impossible to control him. Rockwell hitched Pilot to his wagon and, with a Thoroughbred tied to the back of the wagon, he headed southward. The visit of the peddler's wagon was a big event in the villages of the frontier, for the wagon could provide everything from a darning needle to the latest news, eagerly sought after where there were no newspapers. Nearly every village had a horse the residents believed to be unbeatable. Sometime during the peddler's visit the subject of racing would come up and a match would be made between the village champion and Pilot or the Thoroughbred. The peddler collected a lot of bets, and when Pilot pulled the wagon toward the next village the sporting group watched in stunned disbelief. They were in the same situation as the country boy who lost all of his money betting that Louisville was the biggest city in the world.

During the journey the Thoroughbred died, but Pilot went right on beating the best that could be offered from Philadelphia to New Orleans. In this southern city were rich men who liked to race horses. One asked for a private trial and after seeing the terrific speed of Pilot, a Major Dubois bought the pacer for $10,000, which the accounts say "was paid off in sugar." Pilot then beat the best of New Orleans with such ease that finally he was unable to get a race—no one would start against him. Two men from Louisville named Heinson and Poe were visiting New Orleans and they liked Pilot so much that they paid Major Dubois the same price that he had paid for the horse. Pilot was shipped to Louisville and there he stayed until he died in 1853. The pacing stallion was popular and many mares were sent to him, but his numerous progeny were not noted for their racing ability. The all-purpose saddle horse was an indispensable part of Kentucky life at that time and most of the Pilots went for that purpose.

Although Pilot was a pacer, his son Pilot Jr., which Alexander had acquired, was a trotter and transmitted that gait to his offspring with uniformity. The gray stallion went to Woodburn at the age of four and remained there until the Civil War raids caused his removal to Illinois shortly before his death.

Pilot Jr. not only stood at the head of the Bluegrass stallions but he was the first Kentucky-bred trotting stallion to gain national recognition. Although his colts had speed, it was his daughters that gave Pilot Jr. his far-reaching influence. One of these was Midnight, whose son Jay-Eye-See became the first trotter to cover the mile distance in two minutes and ten seconds. Jay-Eye-See was king for but one day. His record was lowered the next day by Maud S., and she too had a daughter of Pilot Jr. for a mother. Both of these champions came from Woodburn.

The ill-fated Alexander's Abdallah came to Woodburn in 1862. He was the first son of the great Hambletonian to stand in Kentucky; his dam was one of the four mares sent to Hambletonian when he was an unknown youngster. The resulting colt had several names and was first called Edsall's Hamble-

MAUD S., an early trotting champion bred at Woodburn Farm

HAMBLETONIAN, the great father of trotters

tonian. He was purchased by James Miller and J. F. Love of Harrison County, Kentucky, who had gone to New York in search of a horse of the Messenger bloodline. The agreement was that the payment of $2,500 would not be made until the horse was delivered safely. After being driven over the road, the horse arrived in good shape at Cynthiana. He was now called Love's Abdallah.

The Civil War raids on Central Kentucky horse farms made the owners decide to sell their prized stallion, and after four years he was traded to Woodburn for another horse and $2,000. Alexander, who was a British subject, thought he would be immune to the guerrilla raids. He changed the horse's name to Alexander's Abdallah and offered him for service in 1863 and 1864. But Alexander was not as secure as he had thought; Woodburn Farm was raided in October of 1864 and a number of Thoroughbreds were seized, including the stallion Asteroid, which was later ransomed. Alexander was told that more raids were very unlikely as the war was nearing an end, but he was ill advised, for in February of 1865 the trotting section at Woodburn was raided and sixteen of his best animals were taken. A Federal detachment from Lexington engaged in a running battle with the raiders, in which skirmish the sire of good broodmares Bay Chief was killed. Abdallah was not in condition for hard riding as he had been doing heavy stud duty; the raiders abandoned him at Lawrenceburg after sixteen miles. He had been forced to swim an icy stream while in a heated condition, and when found had pneumonia. Although everything possible was done for him, he died four days later.

No one knows the immensity of the loss to Kentucky and the horse world by the death of this thirteen-year-old horse. It was hard to judge his true worth to Kentucky, for many of his Kentucky foals were lost in the war. But while still in New York, Abdallah had begotten a mare called Goldsmith Maid. This mare lowered the world trotting record seven times, the last when she was fourteen, an age when today horses are compelled to retire. The Maid was the darling of the public and raced from coast to coast. When she appeared, the fences surrounding the track area would be torn down by people who wanted to see the champion. She won more races and more money in purses than any other horse of her time, trotter or Thoroughbred. The carefully kept ledger of her driver-trainer, Budd Doble, showed total earnings of $325,000, an amount not surpassed until 1955 by another trotter, Pronto Don. Goldsmith Maid was far ahead of her time. If Alexander's Abdallah could get a daughter like Goldsmith Maid from an old broken-down road mare, he would surely have revolutionized the harness racing sport if he had had access to the great mares at Woodburn and elsewhere in Kentucky.

After suffering heavy losses from the Civil War, Alexander was discouraged and he advertised, "on account of the unsettled condition of Kentucky," a large number of Thoroughbreds and trotters for sale. Fortunately, he reconsidered. In 1864 the Woodburn catalog listed all of the horses at the farm; R. A. Alexander was the first breeder to sell all of his colts instead of racing them. The trotting division had forty young horses listed for sale but thirty-nine young females were reserved to replenish the broodmare band at the farm.

R. A. Alexander died in 1867 at the age of only forty-eight. Had he been spared, the influence of Woodburn would probably have been even greater. As it was, the horsemen flocked to this establishment to get horses to bolster their racing stables and the breeders came to get the great bloodlines to form a foundation for their breeding stock. A brother, Alexander John Alexander, inherited Woodburn. He neither knew nor cared about horse production and left everything to his manager, the able Lucas Brodhead, who had played an important part in helping R. A. Alexander amass his prize collection of the best breeding stock. A. J. Alexander was never in robust health and he was retiring by nature, so visitors seldom saw him and the authority of Brodhead was unquestioned. As Alexander advanced in age, the Woodburn operations were gradually discontinued. Several dispersal sales were held in 1901 and the owner himself died a year later. No other breeding establishment did so much for both the

Dr. Levi Herr, the first advocate of colt racing

Thoroughbreds and the trotters as Woodburn and its influence touches nearly every trotter or pacer racing today.

Another pioneer of the harness sport in Kentucky was Dr. Levi Herr, a veterinarian, who was an apostle of colt racing. Born near Lancaster, Pennsylvania, the same year that the British burned the White House, he arrived in Lexington in 1850 and purchased a tract of 305 acres located about a mile from the Lexington courthouse. He named his farm Forest Park.

Two years after Dr. Herr's arrival Mambrino Chief made his initial appearance at nearby Ashland and the new sire drew the attention of the doctor. Herr always had a partiality for the Mambrino Chiefs, and the great horses bred and raced by the doctor had a considerable influence on the success of that family.

Herr was opposed to the prejudice that many prominent men in the harness sport held against Thoroughbred blood, a prejudice that no doubt was a response to the attitude of the Thoroughbred people toward trotting horses. The doctor realized that the Thoroughbreds had been waiting when the trotters first appeared and that for a while their bloodlines were the only good crosses available for improving the new breed. Although this was the subject of many bitter arguments, Herr proved to be right. The early Thoroughbred crosses did give the boost that started the Kentucky trotters on their way to becoming the elite of the nation.

Herr also bucked popular opinion on the subject of early speed. This was at a time when the general feeling was that "early speed means early decay." Many believed that forcing an immature horse to go at top speed would use up the animal's vitality, cause lameness, and shorten the horse's racing career. Most young horses were not given their first lessons to harness until the three-year-old year. And it took as much as three years to get a horse properly balanced. (Balancing a trotter is finding the best angle of the hoof and the best weight for the steel horseshoes, to encourage a good gait.)

Few were daring enough to oppose these opinions, but Herr had great ability as a breeder and trainer and also a knowledge of veterinary medicine. His method was to begin training as soon as the foal was taken from the side of its dam. These efforts led to the vogue of colt racing. Kentucky owes a debt to Herr, for through his work the state gained the first stallion credited with a family of colt trotters—Mambrino Patchen, a son of Mambrino Chief. Kentucky now stands in a class by itself in the production of winners in all of the noted colt races. The oldest harness race in the world, appropriately named The Lexington, is for two-year-old trotters. It was first raced in 1875 and is still a feature at the Lexington Trots fall race meeting.

3

THE GEORGE WILKES BOOM

THE FIRST SON of Hambletonian to come to Kentucky never had a chance to show his true worth. Had Alexander's Abdallah been spared Kentucky would have started its climb to trotting importance at a much earlier date. It fell to another son of the great father to fan the trotting spark into flame.

George Wilkes was not blessed with a fashionable family tree as he was the son of one of those old-time mares that had no given pedigree. In 1851 James Gilbert of Phelps, New York, happened to notice an attractive mare being ridden by a cattleman on the road between Erie and Meadville, Pennsylvania. A brisk bartering session ensued and the six-year-old mare, Dolly, finally changed owners for $75, with the saddle and bridle included.

Circus owner William Delevan later purchased Dolly in upper New York State for $250. He changed her name to Dolly Spanker, for a character in a popular stage play called *London Assurance*—in spite of the fact that the Dolly Spanker in the play was a man. As she was selected for a circus, Dolly must have been very good-looking. But she was not with the circus very long.

Dolly next went to Harry Felter, a New York grocer and wine merchant, who needed a fast roadster for his business and personal use. He bought Dolly and found that along with her speed the mare had wilful ways. By switching her tail she would get it over the driving reins and then she would run away. To stop this Dolly's tail was cut off short, but it was of no avail for she again ran away and in the ensuing spill her owner's arm was broken. In deep disgust he sent the mare up the Hudson River to his father, who lived near Newburgh, and it was the senior Felter who paid the $25 stud fee when he sent Dolly to Hambletonian in 1855.

The following spring, on a blustery day in March, Dolly Spanker was found dead in the pasture with a newborn foal beside her. The little colt was hurriedly carried to the warm kitchen. The women of the house nursed him and raised him on a diet of cow's milk and Jamaica rum. The youngster was undersized but inherited some of his mother's wilfulness. He would not accept his formula unless a pinch of sugar was added. The colt thrived on this treatment and remained in the dooryard, where he became such a nuisance that he had to be locked up in a paddock, and there he slowly matured. The synthetic diet on which he was raised may have been the reason that George Wilkes remained small in stature.

When the colt started his early training lessons he could show flashes of real speed and this information reached W. L. and Z. E. Simmons, brothers who were well known among the sporting fraternity of the New York metropolitan area. W. L. Simmons went to see the colt and took his trainer, Horace Jones, along. After a trial the three-year-old changed hands for $4,000 and another horse. The new purchase was first raced when he was five, under the name of Robert Fillingham. He won so easily in his first race that he was taken home and carefully prepared for a big race the following year against the champion Ethan Allen. Great excitement was generated. The race drew reporters from all of the large papers. The match developed into a betting race with the former record holder the favorite to beat the young upstart. The Simmons brothers did well with their wagering, as Robert Fillingham won easily in fast time.

After the second year of racing the owners decided to change the horse's name to George Wilkes, in honor of a distinguished friend. Wilkes, the owner and publisher of the turf journal, *Spirit of the Times*, had been among the first to advocate the building of a railroad to the Pacific coast. Later he was decorated by the Czar of Russia for suggesting the trans-Siberia railroad. While covering the Civil War as a reporter, Wilkes contracted a disease that left him a semi-invalid for life. In later years he turned against his friends, including the men who had named a trotting champion in his honor.

Under the name of George Wilkes the horse set the world's record for stallions. He raced twelve years and started in sixty-nine races, of which he won twenty-seven and was second best in twenty-six more. He won over $20,000.

George Wilkes's best record was taken when he was twelve. He was raced long and hard and in his later years was often raced against faster horses. Many times he was brutally whipped in these contests and being a horse of high temper he resented this kind of treatment. He began to sulk and at times he would stop completely when the whip was applied. The racing public is fickle. They had adored George Wilkes in his prime, but they now turned sour toward him; his championship was gone and they were not interested in an uncertain horse.

It had been a case of burning the candle at both ends, for in all of those grueling racing campaigns the stallion had also done heavy stud duty each spring. He had quite a few foals but his offspring could show little speed; the vitality of the sire seemed to have been left on the racecourse. As a result the breeders also turned against George Wilkes.

This great stallion might have slid quietly into oblivion had it not been for William H. Wilson, who had grown up near Chicago and from there had gone to New York. At one time he had been an employee of the Simmons brothers and later became an associate and close friend. Wilson married a Kentucky girl and decided to settle in that state and embark upon a career of raising trotters.

In the 1870s the idea was prevalent that since the trotting breed had been built up in New England, New York, Penn-

George Wilkes

William H. Wilson, who brought George Wilkes to Kentucky in 1873

sylvania, and New Jersey, this section was absolutely supreme in the breed and it would stay that way. There was a slight upturning of the nose as far as Kentucky's early efforts were concerned. The late John Hervey, who was the greatest authority on the American trotter, put this attitude into words: "As for Kentucky, she had sent out only one performer thus far, Lady Thorn, able to compete successfully against the northern stars—and was she not by the former New York stallion Mambrino Chief? Not only was Orange County, N.Y., the cradle of the trotter but his kingdom lay north of the Mason-Dixon line and always would by right of eminent domain. To hold any other opinion was either sheer ignorance or downright heresy!"

Kentucky had long been the hotbed of Thoroughbred breeding and W. H. Wilson was probably the first man to realize that it was also destined to become the great center of trotting horse production. Wilson had a great desire to own the best stallion in the world and the two stallions with the fastest records on the track were located in New York. In 1873 he went there with the purpose of obtaining one of these. After receiving no encouragement from one owner, he then turned to the Simmons brothers and made an arrangement with these gentlemen for bringing George Wilkes to Kentucky. This horse and another he had purchased were then driven as a team over the long distance to their new home. Wilson had great faith in his horse and by his work he raised George Wilkes from the forgotten depths to the heights of success. In doing this he changed the course of trotting history and altered the entire makeup of the breed.

It was a risky venture, as Wilson had only a one-year lease on George Wilkes and he had no farm. He established a stock farm on the Preston place and since it was near the old home of Henry Clay he called it Ashland Park. Many slurs were cast against George Wilkes. One was that no stallion that had raced as hard as George Wilkes would ever amount to anything as a sire. Rival stallioneers were also quick to point out that George Wilkes had sulked. Many of the breeders called him "Bill Simmons's baked-up pony."

The stud fee was set at $100, which was as high as the fees commanded by the more popular and established sires of the neighborhood, but Wilson did one of the greatest selling jobs ever recorded. In his tireless efforts he made deals: if a man brought two or more mares he was given a special rate; some of the better mares were bred on shares; and he even leased good broodmares in an attempt to get foals that would draw quick attention to his sire. His herculean efforts brought eighty-two mares to the court of George Wilkes. But Wilson did not reap the reward that was rightfully his. When Bill Simmons saw the number of mares that had been attracted to George Wilkes he decided that any stallion that popular should be handled by himself alone. Simmons advertised that he was moving to Kentucky to enter the breeding business and wanted to gather a band of broodmares. He took possession of George Wilkes and at first placed him in the charge of A. H. Davenport of Lexington. Later he leased ground about five miles from the city on the Old Frankfort Pike and named it Ash Grove Farm. (The same land was later occupied by E. R. Bradley's Idle Hour Farm.)

The first crop of George Wilkes yearlings in Kentucky were given their first training lessons in 1875 and they all could show speed. Now the prejudice against George Wilkes rapidly melted away. By the time these same colts were racing, their sire had become so popular that owners of the best broodmares were clamoring to have their mares accepted for the formerly despised stallion. At first the Simmons brothers had other stallions of different bloodlines standing as companion sires, but the popularity of the premier sire kept rising faster than the prices on the grocers' shelves and the other stallions were replaced by sons of George Wilkes.

For a horse of advanced age, Wilkes was a remarkably sure foal getter. It also seemed that the move to the Bluegrass had been stimulating, for he was full of vitality and became physically rugged. His virility late in life was unique in the breeding industry. He never lost the sour disposition he had picked up from early rough treatment and would allow no stranger to get near him. He was even more brusque with animals than with people and would attack a dog or cat on sight. The stallion always had a will of his own; at times he would completely ignore mares that were brought to him.

In his nine years in Kentucky, George Wilkes built a great family. One of the remarkable things about him was that after twelve years of hard racing he retired with no leg defects and this resistance to lameness he passed on to his descendants. Another remarkable characteristic was that George Wilkes could either trot or pace and could show speed at both gaits. One of his prominent sons, William L., could also be made to pace when he was shaken up; the man in charge of him would pace him against anybody's horse "for drinks, a twenty-dollar bill, or anything else." This pacing talent was passed on to later generations. Volomite, the first stallion to excel in siring extreme speed, had four lines to George Wilkes in his pedigree, and although himself a trotter is credited with siring twice as many pacers as trotters.

Sometimes a stallion passes on to his descendants some distinctive trait that is noticeable enough to become a sort of hallmark, and this was the case with George Wilkes. The most famous trait of his family was described by the journalist Charles T. Foster as follows: "His hind leg when straightened out in action as he went his best pace reminded me of a duck in swimming." The phrase "the duck stroke of the Wilkeses" could be heard around trotting tracks for many years.

Wilkes produced several important sires in his first crop of foals but he sired some of his most important sons late in life. At twenty-two he sired Guy Wilkes, a horse that became one of California's best sires. One year later Wilkes Boy and Wilton were conceived. The former was a great racehorse and sire but the latter was exported before he had a chance to prove his merit. At twenty-four Wilkes sired Gambetta Wilkes, the sire of the many-time champion George Gano. Two of his best sons, Baron Wilkes and William L., were sired when George Wilkes was twenty-five. The Baron Wilkes line has faded in recent years but William L. carries on. William L. figured in one of the great crosses of bloodlines, called the Golden Cross. This was the mingling of the bloodlines of his grandson Axworthy with those of Peter the Great. Both stallions were located in Kentucky when this great cross was started.

Not all of the sons of George Wilkes carried on with lasting families, but some made great contributions. A good example would be Patchen Wilkes, which was acquired by Henry Jewett for his stock farm just outside of Buffalo, New York. Soon afterward Jewett started an additional farm on an immense tract of prairie land at Chenery, near Wichita, Kansas. The less important broodmares and much of the young stock were shipped to the western annex. Jewett also had a surplus of stallions, having collected several of the sons of George Wilkes at the peak of the boom. Some of them, including Patchen Wilkes, were shipped to Kansas and they practically went into exile. The mares of the Kansas farm harem were not the best of those owned by Jewett and the stallions drew few outside mares from the Kansas breeders. Even so, it was here that Patchen Wilkes sired the great racehorse Joe Patchen, who met the best pacers of his day. He never managed to obtain the world's record for pacers but he did manage to hold the record for a mile over a half-mile track.

After Patchen Wilkes sired his famous son Joe Patchen, the partners Peter Duryea and W. E. D. Stokes decided to buy the stallion. They purchased a farm near Lexington, directly across from Hamburg Place on the Winchester Pike and named it the Patchen Wilkes Farm in honor of their new purchase. But their dreams of speed production did not come up to expectations, for Patchen Wilkes proved to be a dud as a sire. Outside of Joe Patchen he only had one foal with a decent record, Patchen Maid, better known about the tracks as "the diving elk," for her maneuvers while racing.

Patchen Wilkes did not have much speed but he did have a vile temper and was called a "maneater." He was the most savage horse of the entire Wilkes line. It took several husky grooms to get him in harness and out of the stable. As he grew older his temper grew worse and late in life it was necessary to keep him chained in his stall. No one dared enter his stall without some weapon for protection and it was necessary to shove his feed into the stall with a pitchfork.

Patchen Wilkes was finally sold to Tennessee and disappeared from sight. His great contribution was to come years later when as a young horse his son Joe Patchen sired Dan Patch, one of the immortals of the harness sport. Unbeaten in his races, Dan soon found himself without competition as the owners of the other horses refused to start against him. The great horse then started on a series of exhibitions and was eagerly sought after to perform fast miles. Dan became probably the most famous horse in the country. Hobby horses, sleds, cigars, and songs carried the name of Dan Patch.

Even today there is an aura of respect for any Standardbred who can enter the charmed circle of two-minute horses. In Dan's day such a mile was almost a miracle for he did not have the benefit of the highly manicured racecourses nor the advanced equipment used by today's horses. In 1903 Dan Patch paced the Red Mile in 1:59 1/4 and in 1905 in 1:55 1/4. Before he was through Dan Patch had recorded thirty-four miles in two minutes or faster. His claim of being the greatest pacer of the century can best be proved by the fact that no other horse was able to record that many fast miles until Albatross set the new record of thirty-six in 1972.

For a horse that had been cast aside and rejected by the wise people, George Wilkes certainly managed to surmount all difficulties. He put Kentucky on the trotting map and proved that great harness horses could be found south of the Mason-Dixon line. Now the northern breeders and owners began to come to Kentucky in search of top racing and breeding material. To this day these buyers flock to the Lexington sales to bid on the choice yearlings, most of whom have a trace of George Wilkes lurking in their backgrounds.

4

SPREAD OF THE KENTUCKY INFLUENCE

As trotters improved, the world began to beat a pathway to the Bluegrass. Woodburn was not only the first farm to sell all of its yearlings but its selective breeding methods made it the first in the South to produce champions; Woodburn's success drew the attention of the northern horsemen. One of these was William Russell Allen, a prominent New England breeder, who went to Woodburn to obtain the choice stock that became the foundation for his farm near Pittsfield, Massachusetts.

Another visitor to this pioneer Kentucky establishment was Christopher F. Emery of Cleveland, Ohio, who stocked his Forest City Stock Farm liberally with the Woodburn trotters. The Emery estate formed a part of the North Randall mile track, which was prominent in harness racing up to the time of World War II. Emery had good horses and once sent a stable of his trotters to race in Russia.

Dr. Levi Herr had helped to send the old Village Farm at East Aurora, New York, on its way to fame. Mambrino King was the pride of owner C. J. Hamlin. It was said that in his last days Hamlin would sit before a painting of this horse from Kentucky with tears streaming from his eyes.

Dr. A. S. Talbert, a Lexington dentist, loved outdoor life and made his home in the country. He needed good road horses for the daily drives back and forth to his office in Lexington. Talbert's love of farm life included horses, especially trotters, and he made a hobby of breeding these fine animals for his use on the road. The doctor never had a large operation, probably not over ten mares in all, and he did not buy high-priced horses for his venture; his mares were purchased for about $300 each.

Talbert must have been a good horse trader, for one of his deals was the trading of a lot in Lexington, worth about $200, for a beautiful filly; it was one of the greatest broodmare investments of all time. She was given the attractive name of Alma Mater.

Her matronly career began early and she was sent to George Wilkes, for Dr. Talbert was one of the few Kentuckians who really supported "Bill Simmons's baked-up pony" when that horse first appeared in Lexington. At the age of four she had her first foal, Alcantara. He was trained by "Uncle Mike" Bowerman, a former Ohioan who came to Kentucky in 1870. He and his brother were connected with most of the horse deals in the Lexington area. Alcantara set the world's record for four-year-old trotters and was later sold for $12,000 to Elizur Smith of Lee, Massachusetts.

All of the foals of Alma Mater were given a name starting with the first letter of the alphabet. Her second foal was Alcyone. After being raced hard, Alcyone was also sold to Elizur Smith for $20,000. This stallion died at the early age of ten, on the threshold of what promised to be one of the greatest careers a stallion ever had. Alcyone was acclaimed by many as the greatest son of George Wilkes.

After the death of Dr. Talbert in 1883 his son Percy S. Talbert took over the farm and that same year Alma had a foal named Allendorf, who founded a family of Tennessee running walkers. In 1889 Alma Mater was sold to W. S. Hobart, of California, and she disappeared from sight on the West Coast.

The fame of the Kentucky harness horse was spreading. It was a product of Woodburn Farm, Almont, which became the first horse of this section whose foals were to get worldwide attention. Almont, a son of the war casualty Alexander's Abdallah, was to start two important Kentucky harness horse nurseries on their way to fame.

Richard West, a Kentuckian by birth, attended Georgetown College and then chose farming as his occupation. He dealt in cattle and mules on a large scale. In 1868, with the Civil War over and the trotting industry developing rapidly in Kentucky, he decided to become a breeder of trotters.

Over at Woodburn the young Almont was attracting attention. He had raced only once but he won so easily that it was no contest. West approached R. A. Alexander and found that Woodburn was willing to sell, as it had other young stallions of similar bloodlines. The price was $8,000 and Almont went to his new home at Edge Hill, in Scott County. The purchase was deemed so extravagant that even West's closest friends had suspicions concerning his sanity. But the wisdom of West's choice of horses was well verified, for in the seven years that he had Almont this price was paid many times over, not only in stud fees but in the sale of his offspring. Almont, plus the integrity of Richard West, brought Edge Hill Farm to prominence among early postbellum trotting nurseries in Kentucky.

At about this time another farm was being established by General W. T. Withers, a native of Harrison County. He was named for a distant relation, the seventeenth-century English diplomat Sir William Temple. Educated at Bacon College, in Harrodsburg, he was to have delivered the valedictory address but he entered the army before graduation. While leading a charge at Buena Vista, in the Mexican War, he was left as a dying man. He recovered but his wound bothered him all of his life and eventually caused his death. After marrying a Mississippi girl, Withers went to that state and there his law practice, together with cotton planting and partnership in a New Orleans commission house, made him a wealthy man. At the start of the Civil War his fortune was estimated at a quarter of a million dollars. Although he opposed secession, he cast his fortunes with the South and organized a regiment of artillery, the only one of its size in either army. After

General W. T. Withers of Fairlawn Farm

Richard West of Edge Hill Farm

being appointed commander of all the land batteries of the Confederacy, he earned the respect of General Grant while opposing him at Vicksburg. This respect later grew into a friendship. At the time of the final surrender Grant furnished Withers with papers of protection for his family during the period while they were refugees.

The end of the war found Withers penniless. He was appointed as Mississippi agent to Washington, but after returning home he found the disorganized condition of Mississippi made this no place to begin life anew, so he went north to Kentucky. Since his law practice could not give his children the advantages he desired for them, he decided to devote his entire energy to breeding trotting horses. Although the war had left him a poor man, his friends rallied to his assistance, for they had confidence in his judgment.

He purchased a farm and named it Fairlawn. It was then situated in the immediate environs of Lexington but has now been engulfed by the expansion of the city. (The Withers mansion still exists and is now the home of the *Thoroughbred Record* and *The Horseman & Fair World*, a harness horse journal.) Withers selected only the best mares for his broodmare band in 1871, but his first signal success was the purchase of Almont in 1875. He wanted no stallion that had been heavily raced and at that time Almont had not made more than four public appearances. Many people laughed at Withers for paying $15,000 for the stallion but history was to prove that it was a wise investment. It was not long before the family of Almont was sweeping the colt races in Kentucky and as his foals grew older they ventured far afield and won fame all over the nation. Within a decade Fairlawn became the most widely known harness horse stock farm in America and was even forcing Woodburn into the background.

General Withers was the most adroit stallion owner in the business and he revolutionized the system of publicity. He loved to refer to Almont as "the great sire of trotters" and he left no stone unturned to keep his horse in the public eye. He succeeded so well that he became the envy and the despair of the other breeders.

General Withers and family
at Fairlawn

Withers never stooped to anything questionable and he made no false claims. He sold only at private sale and often by correspondence. That the public had faith in his integrity is shown by the number of buyers who were willing to purchase Fairlawn horses sight unseen. The young horses by Almont went to all parts of the United States. General Withers took great pride in his reputation for never misrepresenting any animal he sold.

As the fame of the Almonts grew, the general was able to place high prices on his sales stock and he rapidly amassed another fortune. His sales in some years reached totals that were fabulous for those times.

Under General Withers, Almont achieved a worldwide reputation as a sire of trotters. Among the best foreign customers were Germany, France, and Italy in that order. England, Russia, Denmark, and Sweden were also well represented upon the cashbooks of Fairlawn. The West Indies and the islands of the Pacific were customers. King Kalakahua of Hawaii was a good customer even before he visited America and stayed at Fairlawn.

There were many other visitors at Fairlawn. Among them was U. S. Grant after his presidency; his last visit was just before his death. Old friendship was not the only reason for this visit, for Grant too raised trotters and bought some of his choice stock from Fairlawn.

The broodmare owners fought for bookings to Almont and as a result the stallion was overbred. The great drain on his vitality caused Almont to become almost impotent before his death at the age of twenty. Like so many stallion owners, General Withers had fallen in love with his horse and he had carefully sent the young sons of Almont to buyers mostly outside of the state of Kentucky. He was prompted by jealousy and the fear that one of the sons might overshadow the popularity of Almont. So when the latter died there were none of the Almont family to carry on at Fairlawn. The other stallions that were brought in never attained the popularity of General Withers's "great sire of trotters."

After Withers purchased Almont in 1875, he soon realized the importance of another stallion of superior quality. After careful search and much consideration, he bought from Robert Steel of Philadelphia in 1879 the well-bred Happy Medium, a son of Hambletonian 10. During their years as companion stallions at Fairlawn, Almont's popularity was definitely superior to that of Happy Medium. But Happy Medium sired the great world champion trotting mare, Nancy Hanks, and in 1884 was the leading sire of new standard record performers. Although he sired a world champion, Happy Medium is best remembered as the sire of Pilot Medium, who in turn sired Peter the Great, one of the sport's great foundation sires. Happy Medium died at Fairlawn in 1888, after which the farm and all its horses were sold. Happy Medium is buried on the grounds of the *Thoroughbred Record* and *The Horseman & Fair World.*

In more recent times the greatest establishment to spread the Kentucky trotting influence has been Walnut Hall Farm, a breeding farm unique in many ways. In Europe there are stock farms that have remained in the same families for centuries, but in America the death of an owner generally meant a dispersal sale. The only farm in this hemisphere to continue to raise trotters on a large scale under the ownership of the same family is Walnut Hall Farm.

This historic tract goes back to a grant awarded in 1777 by the state of Virginia. Governor Patrick Henry granted 1,000 acres of land to his brother-in-law, Colonel William Christian, for services in the Revolution. Several owners were connected with this acreage. One was Matthew Flournoy, who came from New Orleans to Lexington to purchase 400 acres of this primeval forest land for a summer home. He built a large brick mansion on the property and the name Walnut Hall came from the native wood with which the interior was finished. The original mansion, which was destroyed by fire, was the birthplace of Sallie Ward, one of the most famous belles of Kentucky antebellum days.

The original tract of 400 acres, including the house built

in 1840 on the site of the first mansion, was purchased by Lamon V. Harkness in 1892. Son of oil millionaire Stephen V. Harkness, Lamon Harkness had gone west to Eureka, Kansas, where he raised cattle for a few years. The death of his father, and a growing family, caused him to head eastward, where he maintained an office in New York City and a home in Connecticut. While in the West, Harkness had had continual use of horses and learned to love them. In order to obtain horses of the quality he desired he had made trips to Kentucky, which delighted him so much that he decided to settle there. The mansion of 1840 was enlarged and remodeled to meet the requirements of a permanent residence. Kentucky had many absentee landlords but Lamon Harkness intended to live at Walnut Hall.

Starting with a few trotters, he first leased a stallion but later used the young stallions born to the mares he had bought to build up his farm. The first of these never raced, as Moko was injured in a stable accident. He started stud duty at three and from his first crop of foals came Fereno, the first to win both the two-year-old and three-year-old divisions of the Kentucky Futurity.* Moko's second Kentucky Futurity winner was the champion filly The Real Lady.

Walnut Hall grew rapidly in both size and prestige. With the purchase of adjacent properties the place grew to over 5,000 acres. It is the oldest successful trotting horse farm in America and at one time had the greatest band of stallions ever collected. Its products won more futurities and important colt races than those of any other breeding establishment. Moko, the first of the great Walnut Hall stallions, was followed by numerous others, including San Francisco, Guy Axworthy (the first to sire four two-minute trotters), Peter Volo, Volomite, Scotland, Protector, Darnley, Guy Abbey, and Kimberly Kid, all two-minute sires.

When L. V. Harkness passed on in 1915 the estate went to

---

* A futurity is a stake in which the dam of the competing animal is nominated either when in foal or during the year of foaling.

The mansion at Walnut Hall Farm

SCOTLAND

his eldest daughter, Lela. Her husband, Dr. Ogden Edwards, took over the management and retained all of the key personnel on the farm. Walnut Hall thrived under his guidance and continued to send out great racehorses. The regime of Dr. Edwards lasted until 1941 and after his death his son, Harkness Edwards, succeeded him as manager. The tenure of the latter was short as "Harky" died prematurely in 1946 and was followed shortly after by his mother.

Walnut Hall then became the joint property of a daughter, Katherine (Mrs. H. W. Nichols, Jr.) and Mrs. Harkness Edwards, who later became Mrs. Sherman Jenney. Though all of the famous stallions were jointly owned, the two owners now had separate operations. Mrs. Nichols retained the name of Walnut Hall Farm and Mrs. Jenney operated Walnut Hall Stud. The huge broodmare band was divided between them and both maintained the high standards of the old nursery. The premature death of their sire Rodney in 1963 caused the Jenneys to lose interest and Walnut Hall Stud was dispersed. In 1972 Walnut Hall Stud was purchased by the State of Kentucky and is now the Kentucky State Horse Park. Walnut Hall Farm continues in the great tradition dating back to 1892.

Walnut Hall was the first farm to send all its yearlings through the sales ring to the lilting chant of the auctioneer. Starting in 1906, all yearlings were auctioned regardless of depressions, wars, or whether the market was high or low. At first some Walnut Hall yearlings were sent to Chicago and the rest to Old Glory sale in New York, operated by the Fasig-Tipton organization. (Partner Ed A. Tipton always used to say that he was just a "country boy from Kentucky.")

Walnut Hall brought about the demise of the Old Glory sales company, for in 1938 the Kentucky nursery decided to sell its yearlings at home during the Grand Circuit meeting at the Red Mile. The decision to sell the yearlings at the farm went back to the early days in Kentucky when a breeder usually sold from his base of operation. Without the great consignment from Walnut Hall, Old Glory no longer drew the buyers with big money and the sale quietly slipped into oblivion the following year. Fasig-Tipton, however, remains active in conducting Thoroughbred sales.

The horses from Walnut Hall have long had a worldwide reputation. For many years foreign buyers were always present at the farm's sales and the products from this nursery have spread all over the world. Most of the European countries can be listed among the importers. Many Walnut Hall products have gone north to Canada, some to South America, and some even to Japan. Australia, New Zealand, and Tasmania have been heavy buyers over a long period and the Walnut Hall strain imported by these countries has been an important factor in establishing the bloodlines of the important horses Down Under.

Several sales companies were located in Lexington in the nineteenth century. Two of the earliest of these were headed by W. R. Woodward and Wiley Brasfield. For a period of several years before 1892, horse sales were booming and a British Thoroughbred organization called Tattersalls that year decided to move into this lucrative American business. Tattersalls set up three trotting horse sales, one in New York, one in Cleveland, and one in Lexington. But it was the wrong time for the English concern to enter this business, for with the panic of 1893 prices hit rock bottom. The name Tattersalls is still used in Lexington but there have been several ownerships since the British-owned Tattersalls sold out. The Lexington sale carried on in a small way and for many years held combination sales of harness horses of all ages. Sometimes one of the farms would even send in a few saddle horses to be sold. But Walnut Hall Farm was indirectly to change the history of this sale.

The Lexington races each fall drew people from all over the world and among these were some who paid good prices for the Walnut Hall offerings at the farm sales. Leo C. McNamara of Two Gaits Farm at Indianapolis noticed this and in 1942 he sent in a choice consignment of yearlings to Tattersalls. This was the start of the present greatness of Tattersalls, for other farms soon followed suit: Castleton, Almahurst, Gainesway, and others began to patronize the

Lexington sale with their best stock. In spite of losing its sales arena twice by fire, Tattersalls steadily grew and now even Walnut Hall sends its yearlings to this sales arena.

During the October races in 1976 Tattersalls had a banner year, for the yearling sale brought the largest total—over $8.5 million—in harness horse history. The largest average price for a large consignment from any trotting horse nursery was posted by Stoner Creek Stud of Paris, Kentucky, in that same sale.

Tattersalls also holds the record price for a yearling sold at auction. In 1971 Stoner Creek Stud sold Good Humor Man for $210,000. Good Humor Man took a fast record and was sold for export to go to the Antipodes. In the 1974 sale for horses of all ages, Delmonica Hanover sold for $300,000 to give Tattersalls the credit for the highest all-time auction price for a harness horse.

Kentucky has come a long way since Woodburn Farm first decided to sell all of its yearlings.

5

FOUR FAMILIES OF TROTTING

LIKE THEIR HUMAN COUNTERPARTS, horse families are named for the masculine side of the pedigree. Over the span of many years four stallions emerged as the patriarchs of trotting clans. All four were either born in Kentucky or had their taproots deeply embedded there. Two of them made their reputation as great sires while in the Bluegrass.

PETER THE GREAT. The dominant trotting family of today has stories in its background that read almost like fairy tales. One of the early branches came from the plantation of T. J. Wells of Rapides Parish, Louisiana. This family had lived under five flags, as Connecticut was still a British colony when the father was born. After migrating to Louisiana they lived under the banners of France and Spain, under the stars and stripes, and also under the stars and bars of the Confederacy.

General Wells had a Thoroughbred mare that he used as a utility horse because an attack of distemper had left her unfit for racing. A mistake by a slave named Ike Dixon caused the mare to be bred to a stallion, Aikenhead, which could neither trot nor gallop, but did pace.

When Scott's Louisiana Cavalry was being organized, a nephew of General Wells was given the mare for his mount and no one was aware that she was then carrying a foal.

- PETER THE GREAT
  - Pilot Medium
    - Happy Medium
      - Hambletonian
        - Abdallah
        - Chas. Kent Mare
      - Princess
        - Andrus' Hambletonian
        - Unnamed
    - Tackey
      - Pilot Jr.
        - Pilot
        - Nancy Pope
      - Jenny Lind
        - Bellefounder
        - Unknown
  - Santos
    - Grand Sentinel
      - Sentinel
        - Hambletonian
        - Lady Patriot
      - Maid of Lexington
        - Mambrino Pilot
        - Brown Lock
    - *Shadow (Lady Duncan)
      - *Octoroon Jr. (Saddlerville)
        - Octoroon
        - Unnamed
      - *Swallow (Dixie)
        - Creole
        - Bettie Wilson

* Known under two names

Young Wells named her Lady Bess and rode her until she became so heavy that his fellow troopers begged him to trade her off as a safety measure. After the fall of Fort Donelson, in February of 1862, Company D was retreating in snow a foot deep. At Clarksville, Tennessee, it was necessary to dispose of Lady Bess.

She was traded off for a bald-faced gelding from the lot of Joe Thomas, the village blacksmith. A few weeks later Lady Bess dropped a black foal that was later named Creole and used for a saddle horse. As a four-year-old he sired a daughter named Dixie. That fall Creole was destroyed by fire, but Dixie lived to make some Standardbred history.

Some time before that, a man from Lexington, while on a riverboat going to Louisville, got into a poker game with a gambler named James Madison. After going broke the Lexingtonian put up a horse called Octoroon and lost him too. Octoroon was raced continually until he was accidentally shot while running in a pasture. Before his death he sired Octoroon Jr., one of the most versatile horses even seen. "Pop" Geers, one of the greatest harness horse drivers at the turn of the century, once saw Junior win three races in one day—one on the trot, one on the pace, and one over the hurdles.

A daughter of Octoroon Jr. and Dixie, given the name of Lady Duncan, could show speed on the trotting gait. When S. A. Browne was selecting broodmares for his farm at Pentwater, Michigan, for some unknown reason he visited Nashville, Tennessee, and paid $3,000 to the banker William E. Duncan for this totally unknown mare. He changed her name to Shadow. After one season of racing the new owner was convinced that Shadow was useless for racing purposes, so he used her as a road mare. Browne drove her between his lumber camps in upper Michigan and she became the most noted road horse of that section. He also sent her to a grandson of Hambletonian, Grand Sentinel, which was standing at Lexington.

Grand Sentinel was a gentle, clever horse and could always be driven with a loose rein. In the cold and icy January of 1887, Grand Sentinel was turned out in his paddock for exercise. While playing he slipped on an icy spot and took a hard fall, which hurt him so badly that he was unable to get to his feet and died soon afterward, at the age of thirteen. Shadow's daughter Santos was a posthumous foal of Grand Sentinel.

As a yearling Santos was purchased at the Cleveland sale for $850 by J. I. Case and one year later she went through the Brasfield sale at Lexington. A bid of $470 sent the filly to H. D. McKinney, Case's brother-in-law, who was from Janesville, Wisconsin. This man traded the mare back to Browne, who thought Santos of so little value that he promptly sold her to D. D. Streeter of Kalamazoo, Michigan. It was Streeter who sent Santos to Pilot Medium, a crippled gray stallion from Pennsylvania that belonged to Walter Clark of Battle Creek, Michigan. The son of Pilot Medium and Santos was Peter the Great, who was to become one of the greatest patriarchs of trotting.

The paternal side of Peter's pedigree goes back to New Orleans. Jenny Lind, named for the famous "Swedish nightingale," had been taken there to race and became a permanent cripple after a stall accident. She was sold for a low price to a man by the name of McHatton, lessee of the prison at Baton Rouge. He sent Jenny to Woodburn Farm and while there she had three daughters by old Pilot Jr. One of these died but the other two were returned home and turned out on the prison commons to shift for themselves. The prisoners named one of them Tackey, a southern term that meant "a scrubby, neglected, unkempt horse raised on its own." Tackey passed through many hands and was raced hard. When she was fourteen her name was changed to Polly. She closed out her racing career in Pennsylvania, her name was then changed back to Tackey, and she was bred to the stallion Happy Medium, who at that time was still owned by Robert Steel. The resulting foal was the gray cripple, Pilot Medium, who became the sire of Peter the Great.

Peter was not a natural-gaited trotter like most of the Pilot Medium family. It took all year to get him ready for his first race and he was second in the junior division of the Kentucky Futurity. The following year he trained so badly that he was kept eligible to only one race, but he became the first northern colt to win the senior division of the Kentucky Futurity. Peter the Great set a new record for the race, winning by twenty lengths in 2:12 1/2 in 1898.

Although the effect of the panic of 1893 was still felt at that time in the prices of horses, J. Malcolm Forbes of Boston, Massachusetts, paid $15,000 plus $5,000 worth of other considerations for Peter the Great. The horse turned out to be a bitter disappointment to Forbes, for he was unable to get speed while he was driving the stallion. Forbes grew to hate the animal and would not even allow him in the barn with the other stallions at Forbes Farm.

From the few mares sent to Peter the Great he sired Sadie Mac, unbeaten in her races against the best trotters of the nation. She lost her only race the hard way, dropping dead during the contest. The fame of this mare caused her sire to bring $5,000 at auction. The buyer was Peter Duryea of New York City, the partner of W. E. D. Stokes in Patchen Wilkes Farm at Lexington. The purchase caused a disagreement between the two and Stokes would not allow the horse on the farm. But an offer of $8,000 from other buyers caused Stokes to change his mind. Peter the Great, while standing at Lexington, went on to become the greatest sire of his time. After the horse became famous, Stokes claimed that he had personally selected him.

In 1917, when Peter the Great was twenty years old, Stoughton Fletcher paid $5,000 and took the great progenitor to Laurel Hall Farm at Indianapolis, where the horse lived out his life. For many years after his death Peter the Great lay in an unmarked plot; the American Legion erected a monument to him in 1938.

The foals of Peter the Great were sensational and in such demand that they rapidly spread over the nation. In one futurity over a hundred foals by Peter the Great were entered. Because of the lack of records at Patchen Wilkes Farm, no one knows how many foals were sired each year, but there were many.

The sons of Peter the Great carried on in speed production

and were the most famous of their day. Today few pedigrees of good Standardbreds fail to show Peter the Great as one of the ancestors and he is generally found several times. Peter the Great was the first immortal voted into the Hall of Fame of the Trotter in Goshen, New York.

AXWORTHY. "Throughout the entire range of horse history, no series of episodes is at once so striking, so unique, and of such enduring interest as those in which C. W. Williams and his two stallions Axtell and Axworthy played the leading roles," wrote John Hervey in *The American Trotter.*

C. W. Williams, born in Chatham, New York, in 1856, was taken as a boy to Buchanan County, Iowa, where when not busy on his father's farm he hired out to neighbors for twenty-five cents a day. Later the family moved to the town of Jesup, where young Williams was employed as a store clerk for five dollars a month plus board and the privilege of sleeping under the counter. After losing that job in the panic of 1873, Williams went to Chicago, where he drove a milk wagon and managed to save some of his twenty-dollar-a-month wages. Then he returned home, to high school and a night job at the railroad station, where he picked up telegraphy. After graduation, Williams started a business of his own, shipping Iowa butter and eggs to New York. He also became the night telegraph operator at Independence, Iowa, a few miles from Jesup, married, and settled at Independence.

Williams had been interested in horses all his life, and he now became a student of trotting pedigrees. In 1883 he bought four young horses from H. L. and F. D. Stout at their Highlands Farm near Dubuque. Two mares in this group were to make Williams a wealthy man. Gussie Wilkes was a cripple, so he got her for $75. Although not fashionably bred, Lou cost $125. This mare had a growth on her hind legs and she later transmitted this to her descendants.

The "Wilkes craze" was then prevalent, so Williams shipped the two mares to Ash Grove Farm, at Lexington, in charge of a trusted employee, John Hussey. George Wilkes was dead by this time, but W. L. Simmons had retained some of the sons of his great sire. The stallions Hussey originally chose were Red Wilkes and Baron Wilkes, but they were well patronized and the stallioneers were not interested in the two cheap mares from Iowa. Williams was always tight-fisted in money matters, probably as a result of the struggles of his youth; he finally agreed to a deal his man made with Ash Grove Farm for the services of two George Wilkes stallions less popular than his first choices. One was Jay Bird, a roan horse who was going blind; the other was William L., a handsome, well-bred young horse with crooked hind legs, who was being offered for his first season. The fee for Jay Bird was $100, with $50 for William L., and as there were two mares, these fees were reduced. Gussie Wilkes was sent to Jay Bird and Lou to the young stallion.

The foal of Lou was Axtell who, after lowering the three-year-old record, was sold for $105,000, which up to that date was the highest price ever paid for a horse of any age, breed, sex, or gait, anywhere in the world. But the bad hind legs of Lou and William L. were passed on to him and lameness was to end the brilliant career of Axtell after he set the stallion time record.

With the money from Axtell, Williams built a farm, a street-car line, and a kite-shaped track at his Iowa home. Many believed that horses could go faster over this racing strip shaped like a figure eight than they could on a regulation oval track. This belief caused horses to be shipped in from all over the nation and finally it was impossible to get accommodations in Independence for either man or beast. Most of the horses were not there to race but to go for speed records. These attempts in which a horse races against a certain designated time are now called "time trials."

The panic of 1893 wiped all of this out. Independence lost its popularity and Williams lost everything except Allerton, the foal of Gussie Wilkes; through Allerton he made another fortune. When the deal was completed on Axtell, Williams had said that he was keeping the better colt. In fact, at one time he refused an offer of $150,000 for Allerton. Although Allerton once led the nation as a sire of speed, his fame did not last because he failed to establish a male line to carry

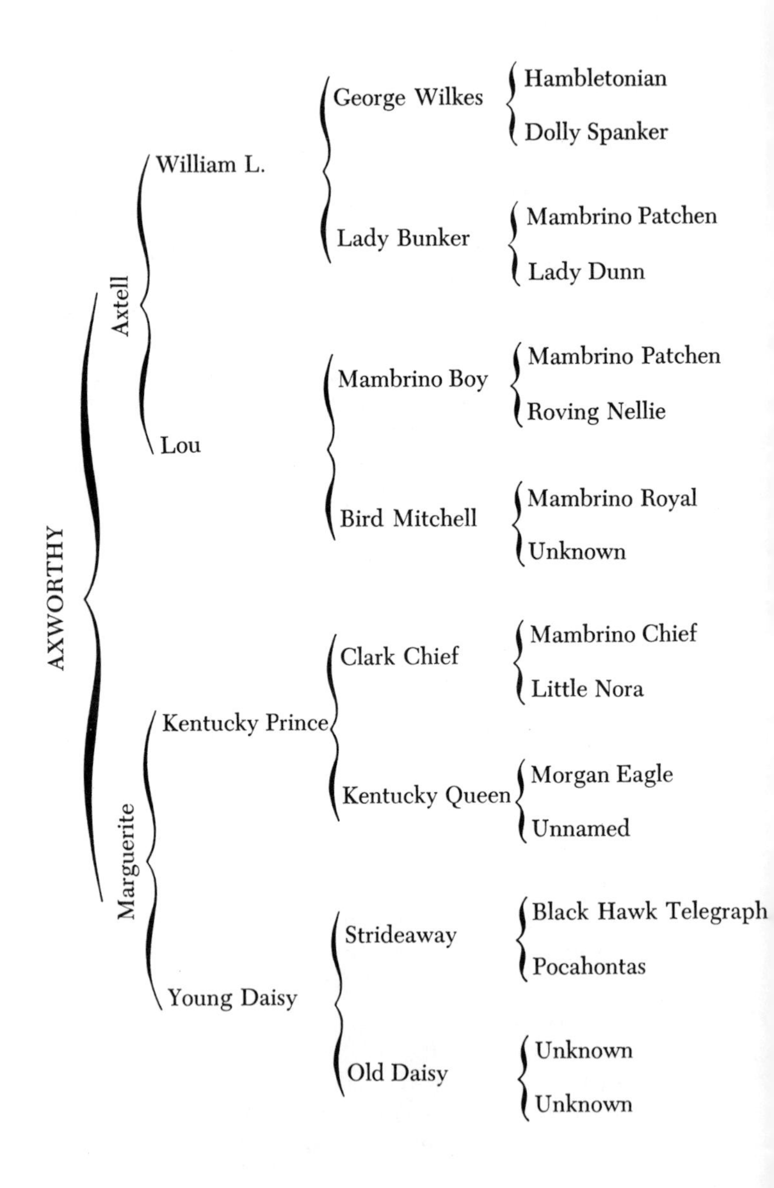
AXWORTHY
Axtell
Marguerite
William L.
Lou
Kentucky Prince
Young Daisy
George Wilkes
Lady Bunker
Mambrino Boy
Bird Mitchell
Clark Chief
Kentucky Queen
Strideaway
Old Daisy
Hambletonian
Dolly Spanker
Mambrino Patchen
Lady Dunn
Mambrino Patchen
Roving Nellie
Mambrino Royal
Unknown
Mambrino Chief
Little Nora
Morgan Eagle
Unnamed
Black Hawk Telegraph
Pocahontas
Unknown
Unknown

PETER THE GREAT

AXWORTHY

on. It was Axtell who was destined to make history. Through his son Axworthy have come such champions as Greyhound and Titan Hanover.

After changing ownership, Axtell again lowered the three-year-old record and became the first of his age to hold the record for trotting stallions of all ages. No other three-year-old was ever able to accomplish this feat. The syndicate that had bought him was headed by the same John W. Conley who did so much for Kentucky when he arranged the lease of Dictator to Richard West. The Axtell syndicate was anxious to get a return on their big investment, so the horse was offered for stud service at a $1,000 fee, double that of St. Simon, then the most popular Thoroughbred in the world. Axtell's fee was the same for three years and when the panic of 1893 came along Axtell had paid for himself.

Conley, in his attempt to get mares for Axtell, contacted A. B. Darling, proprietor of the elegant Fifth Avenue Hotel in New York City and a member of the famed "Sealskin brigade," a group of prominent road drivers in the metropolitan area. Darling had once sent a favorite mare called Old Daisy, of unknown lineage, to the very fast racehorse Strideaway. The 1870 foal of this pair was a gray mare named Young Daisy, whose daughter Marguerite was to be the dam of Axworthy. Marguerite's sire was Kentucky Prince, who had been sent to the farm at Darlington, New Jersey, after an attack of influenza had ruined his chance for a racing career.

In 1878, Darling consigned Kentucky Prince to the combination sale held by Peter C. Kellogg at the American Institute in New York City. This sale was a historic event, being the first of its kind ever attempted. Before this time auctions of fine trotters involved horses from a single owner or stable and such sales had never lasted longer than a day. Kellogg's two-day sale included about a hundred animals from many different owners. His extensive advertisement of the event was criticized, but probably helped him achieve a rousing success. The average price of the ninety-five horses sold broke all such records and Kentucky Prince brought $10,700, then the highest price ever paid for a trotter at public sale. The buyer, Charles Backman, took Kentucky Prince to Stony Ford Farm near Goshen, New York—the first of the descendants of Mambrino Chief to be reintroduced into the section whence the Chief had come twenty-six years before. A. B. Darling sold most of his horses at that sale but he retained his favorite, Old Daisy, as well as her daughter Young Daisy and the latter's two-year-old filly by Kentucky Prince, Marguerite.

When Conley contacted Darling, the latter decided to breed Marguerite, who was by then fourteen, to Axtell. She was left at the farm of W. P. Ijams at Terre Haute, Indiana, where the syndicate was standing Axtell. Although Axtell had done wonders in a financial way for his breeder as well as his owners, Ijams was not impressed. In the final stages of Axtell's career Ijams once remarked, "I think more of my Jersey bull than I do of Axtell."

The second foal of Marguerite was Axworthy. This son of Axtell also had trouble with lameness and he took his race record of 2:15 1/2 in a special race against two colts it was known he could beat. Shortly afterward, A. B. Darling died and his horses were dispersed at a three-day auction conducted by Kellogg in New York. The year was 1896 and the country was in a depression. Trotting horse values had dropped to the point that many animals could scarcely be sold at any price. The auction that day was going poorly, the small crowd offering only the lowest bids. When Axworthy entered the ring no one would bid at all, for one foreleg was obviously unsound and the horse was known to be unable to stand training. Kellogg pleaded for bids to no avail and Axworthy was about to be led out of the sales ring, probably destined to be castrated and used as a buggy horse. At this point John H. Shults of Brooklyn joined the crowd. Shults could not stand to see the horses of his old friend "butchered," as he put it, and his opening bid of $500 got Axworthy. The first thing Shults said was, "What am I going to do with him?" As it turned out, he used Axworthy for stud duty in a small way at his Parkville Farm in the Brooklyn suburbs. Axworthy's first foals came in all sizes, shapes, and colors—but they could all trot.

The family of Axworthy did so well that when the horses of John Shults were moved to Shultshurst, in Westchester County, the owner sold his other stallion and kept Axworthy. When advanced age caused the retirement of John Shults there was no trouble getting bids on the horse at the dispersal sale. The bid of $21,000 by William Simpson, of New York, took Axworthy to the Empire City Stud at Cuba, New York. Here was a unique situation, for Simpson owned two of the family founders at the same time. The other was McKinney, a good-looking horse with a nice disposition, so the owner fell in love with him and this horse was kept at home so that Simpson could see him. Axworthy was a more rugged character, so he was exiled to Kentucky. This was a blessing in disguise, for here he had access to the best mares and he lived out his life under the management of R. L. Nash at Lexington.

The foal that lifted Axworthy to his highest fame was Hamburg Belle. Bred by a former friend of Shults, the mare was given to John E. Madden, of Hamburg Place, to manage. The little mare could show such speed that Madden purchased her and changed her name from Sally Simmons II to Hamburg Belle. After setting the race record at North Randall, Ohio, she was sold to W. M. Hanna, of Cleveland, for $50,000 before she had been cooled out. After being beaten in her next race the mare contracted pneumonia while being shipped south and soon died. Today she rests in the beautiful little horse graveyard at Hamburg Place, opposite Patchen Wilkes Farm.

Axworthy had two great sons—Guy Axworthy, the futurity colt sire and the first to sire four two-minute trotters, and Dillon Axworthy, a great sire of colt speed. The daughters of this family seemed to produce more speed when crossed to the Peter the Great blood. Thus another great cross of bloodlines originated in Kentucky and it was called the Golden Cross: the mingling of the lines of Peter the Great and Axworthy.

BINGEN. May King, the sire of Bingen, had a checkered career. This product of Palo Alto Farm, in California, was foaled in 1886 and sold when two to Miller and Sibley of Prospect Hill Farm in Pennsylvania. The partners in Prospect Hill had become millionaires from the coal, oil, and iron of western Pennsylvania and owned huge tracts of land there. At their farm they first started with Jersey cattle and later became interested in trotters. Among their early purchases was May King, but he was left in California to get a record. While there he was sold for $7,500 to William E. Spier, with the expectation that he would lower the trotting record for two-year-old stallions. Learning that Spier was disappointed in this, and also in May King's preference for the pacing gait, Miller and Sibley accepted the return of the stallion. May King was later sold, for $10,000 and several broodmares, to Smith McCann, of Lexington, who had bought Fairlawn Farm after the death of General Withers. Later, in an attempt to avert a financial crisis, part of the stud of Smith McCann was sent to the Lexington auction in the fall of 1892. In this group was a mare called Young Miss, then heavy in foal to May King. A Lexington physician, Dr. David Bennett, was the buyer and Bingen was foaled his property.

When it was time for training to start, twelve-year-old Raymond Snedeker was sent to get the colt. It was quite a task, for the colt was at the side of his dam. He had never been weaned and had never had a halter on his head. It was a tussle but the boy managed to get the colt back to the Lexington track. Young Snedeker was already helping his father in the training chores and he pleaded until he was allowed to give the colt his first lessons to harness and cart. At this time Dr. Bennett had priced Bingen at $150, but no buyers were interested in a colt that seemed to do everything but trot. Suddenly the little fellow found himself and started to improve rapidly with a smooth trotting gait. Local papers noticed the boy and his colt and the pair were written up in the turf journals. So Bingen was sold to George W. Leavitt, of Boston, for $1,500. Leavitt then sold a half interest to E. H. Greeley of Ellsworth, Maine, for $1,250.

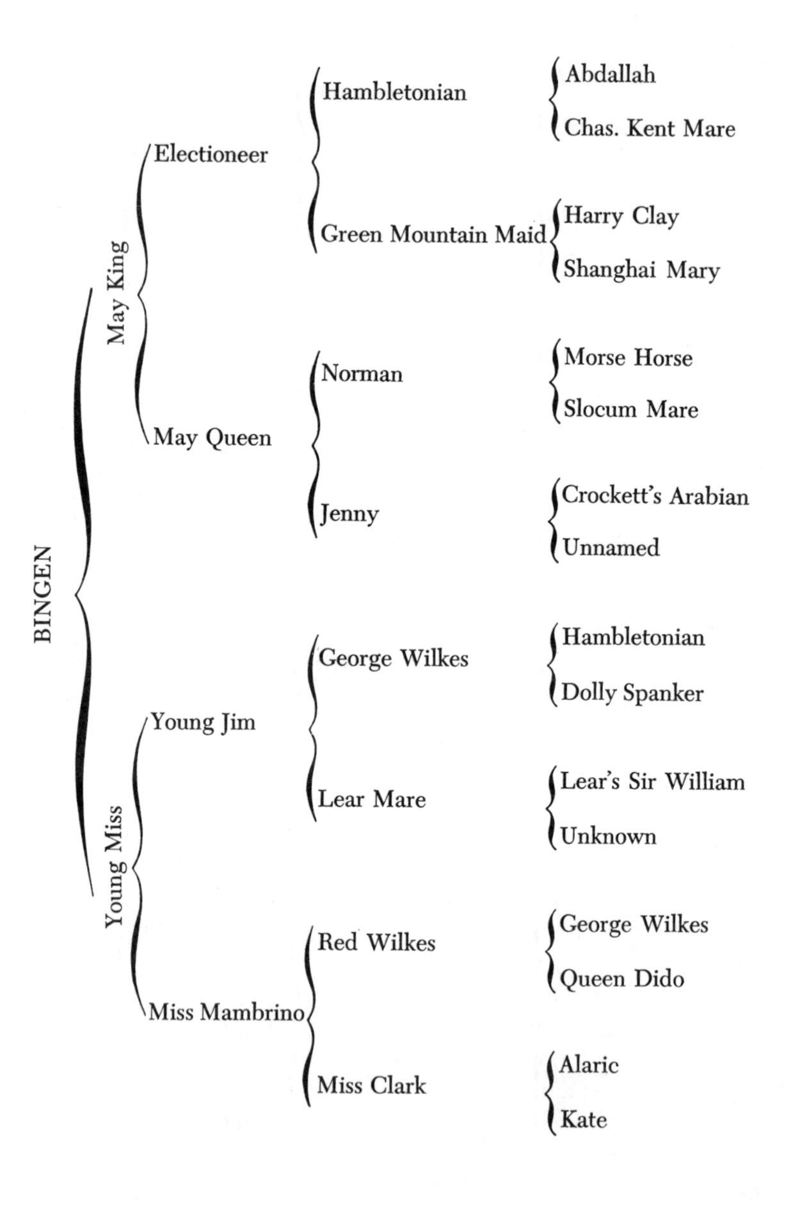

BINGEN
May King
Electioneer
Hambletonian
Abdallah
Chas. Kent Mare
Green Mountain Maid
Harry Clay
Shanghai Mary
May Queen
Norman
Morse Horse
Slocum Mare
Jenny
Crockett's Arabian
Unnamed
Young Miss
Young Jim
George Wilkes
Hambletonian
Dolly Spanker
Lear Mare
Lear's Sir William
Unknown
Miss Mambrino
Red Wilkes
George Wilkes
Queen Dido
Miss Clark
Alaric
Kate

Snedeker, who became one of the leading reinsmen of his time, wrote as follows of his boyhood experience with Bingen:

> I had often worked out horses for my father, but never had any certain one given me to train exclusively by myself. I was anxious to get to work on him, so started right in to break him to harness the next day. After a few days' ground driving he was hitched to a cart and driven once around the mile track. He appeared to be broken the first time hitched, and never gave me any trouble to waywise. But he had every gait but the trot—could not do a thing but gallop, pace, rack, and single-foot, and was unable to take more than two or three steps in succession at any of those gaits. This was before he had shoes on. I shod him first with plain 8-ounce shoes forward, and he straightened right out into a good square trotter. The first time he ever struck a trot he went an eighth in 25 seconds, and every time I stepped him from that time on he could do a little more than in the previous workout. I used quarter-boots on him when I brushed him, but do not remember that he ever hit himself in any place. I used no boots on him and no toe-weights. The trainers at the Lexington track that season said he was the most remarkably level-headed colt they ever saw. After his first set of shoes were put on him I never knew him to break [stride] but once in his yearling form, and that was in his first race, which was trotted over the Lexington track in July.

As a two-year-old Bingen was a sensation and J. Malcolm Forbes paid $8,000 for him. He was raced against aged horses the following year and did not do so well. But at Lexington he won the Kentucky Stake and in this he beat the winner of the Kentucky Futurity. At that time it was thought that the four-year-old year was a bad year because the animal was passing from colthood to maturity. For this reason Bingen was used only in the show ring that year. He won the four classes in which he appeared and wound up with the grand championship for trotting stallions. After resuming his racing career he either broke, equaled, or forced another horse to break, half a dozen records. Later Forbes used him on the Charles River Speedway in Boston. Unlike Peter the Great,

Bingen became the favorite of Forbes and eventually was champion of the speedway.

After the death of Forbes, Bingen had several owners and stood mostly in New England, a section which was known as "the graveyard of sires," but here he founded a family. Later William Bradley used him as a private sire at his Ardmaer Farm at Raritan, New Jersey. When this establishment was dispersed David M. Look purchased Bingen for his newly acquired Castleton Farm at Lexington.

This was to have been Bingen's first real chance, as he now had access to Castleton's great mares, but he died just after he had started his second season at Castleton. Bingen was the only one of the four family heads to sire a two-minute trotter—Uhlan 1:58. Many of Bingen's get were exported, and one of these Codero, won the championship of Europe three times. Several of Bingen's sons were important sires.

Before his death, Bingen and his family were the victims of a bitter, vicious campaign of slander. In describing this John Hervey once said, "It was launched by no one man but rather was the result of a movement formed by a clique inspired very obviously by envy, jealousy, commercial and sporting rivalry, and kindred motives. . . . Beginning quietly, it was artfully sown broadcast throughout the trotting world until the entire atmosphere was vitiated with it. . . . A scandal, even when fictitious, will often outweigh a lifetime of good deeds."

If Bingen had a fault it was the passing on of beautifully gaited early speed to his foals. Many of them were ruined when young by trainers who called on this speed too often, causing lameness and bad manners. Although not one of the dominant families today, the Bingens did contribute hugely to the greatness of other families. Many champions and great horses carry a strain of this allegedly tainted blood.

Of the hate campaign against the Bingens, John Hervey concluded: "It is an object lesson in the harm, injury, and loss which prejudice and antagonism, carried to the extreme, can produce in the breeding industry, as in other spheres of energy."

McKINNEY. One of Dr. A. S. Talbert's inexpensive mares played a stellar role in founding the fourth family. During the third season that George Wilkes was in Lexington the doctor had sent Alma Mater to him and was so pleased with the resulting Alcantara that he returned the mare the following season and got Alcyone. Alcyone showed even more promise than his champion brother, but he was nearly ruined early in his training by a strain across the back that roughened his gait. To add to his troubles the horse was leased to Frank Van Ness, who started him in sixteen races, in many of which he was in no condition to start.

About this time Talbert died and Elizur Smith again entered the picture. He paid the widow $20,000 and took Alcyone back to stand with his brother at Highlawn Farm in the Berkshires. One of the conditions of the sale was that Smith was to honor all bookings the Kentucky breeders had made for their mares. One of these breeders was W. H. Wilson of Cynthiana, who had brought George Wilkes to Kentucky, and when the carload of Kentucky mares was sent to Massachusetts he sent Rosa Sprague. Her foal, McKinney, was born at Wilson's Abdallah Park in the spring of 1887.

It is not known exactly what characteristics of either horse Wilson might have had in mind when he decided to have Rosa Sprague bred to Alcyone—a decision that meant shipping the five-year-old mare a thousand miles to a stud who was closely related to his own stallion Simmons. He had already bred Rosa Sprague to Simmons as a three-year-old, and the resulting filly, Hettie Case, later became the dam of Fereno, winner of the 1900 Kentucky Futurity. After being bred to Alcyone in 1886, Rosa Sprague dropped a seal-brown colt with white hind coronets, named McKinney. As a two-year-old McKinney was purchased from Wilson at the relatively fancy price of $1,500 by Charles A. Durfee, a young Irishman living in Los Angeles. He remained in Durfee's California stable for thirteen years.

McKinney was outstanding on the racecourse. He was unbeaten at four and took his record of 2:11 1/4 in a historic eight-heat contest in his final campaign. He won seventeen

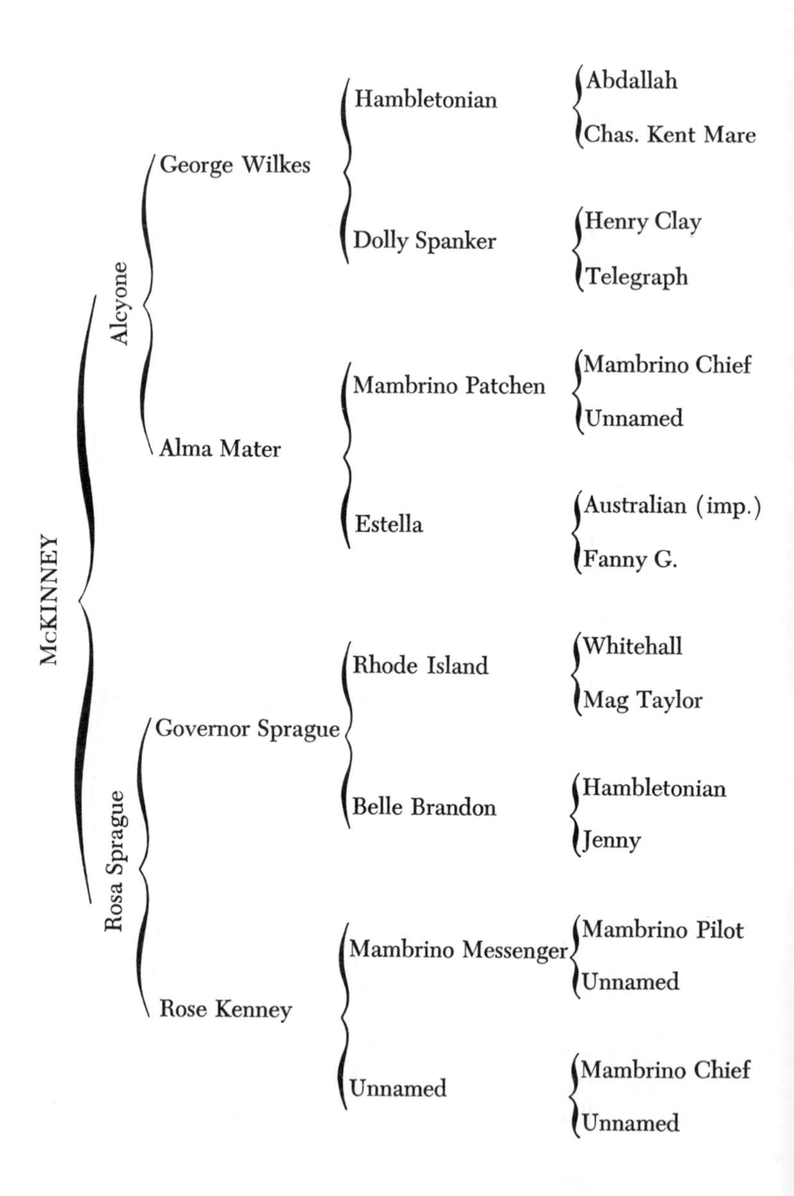
McKINNEY
Alcyone
Rosa Sprague
George Wilkes
Alma Mater
Governor Sprague
Rose Kenney
Hambletonian
Dolly Spanker
Mambrino Patchen
Estella
Rhode Island
Belle Brandon
Mambrino Messenger
Unnamed
Abdallah
Chas. Kent Mare
Henry Clay
Telegraph
Mambrino Chief
Unnamed
Australian (imp.)
Fanny G.
Whitehall
Mag Taylor
Hambletonian
Jenny
Mambrino Pilot
Unnamed
Mambrino Chief
Unnamed

out of his twenty-four starts and this was remarkable as he was doing heavy stud duty each year. The demand for the services of McKinney was so great that he was moved up and down the West Coast and sometimes used as a sire for nine months of the year. After refusing countless offers, Durfee finally sold McKinney for $25,000 to Henry B. Gentry of Bloomington, Indiana, the proprietor of a well-known circus and menagerie. In the thirteen years under Durfee's ownership the horse had earned over $150,000 in race purses and stud fees.

McKinney stood at the Gentry farm for two years. Then a daughter named Sweet Marie swept through the Grand Circuit, the elite of the racing groups. After the mare won the famous old Transylvania Stake at Lexington in 1904, it was announced to the spectators that William Simpson had just paid $50,000 for the sire of the winner. McKinney went to the Empire City Stud at Cuba, New York, where he spent the rest of his life. He died in 1917, aged thirty.

Simpson had a great affection for McKinney, so when he sent his other stallion, Axworthy, to Lexington, he kept McKinney at home. This made McKinney practically a private stallion with only a few mares and no chance to enlarge his opportunities. But McKinney did have two great sons, Zombro and Belwin. The former was conceived while McKinney was in California. He was one of the most prominent of the colts on the West Coast, for he was racing against aged horses. Although many buyers tried to get Zombro, his breeder, George T. Becker, put such a prohibitive price on him that Zombro remained his property.

San Francisco, a son of Zombro, came east and after two seasons his owner, P. W. Hodges, had financial difficulties, so Walnut Hall bought the beautiful bay stallion. San Francisco's chances for contributing to the fame of the McKinney line were wrecked when his three greatest offspring died during their racing careers. They were St. Frisco, one of the greatest racing stallions; Fireglow, who died mysteriously after beating many great colts that later became leading sires;

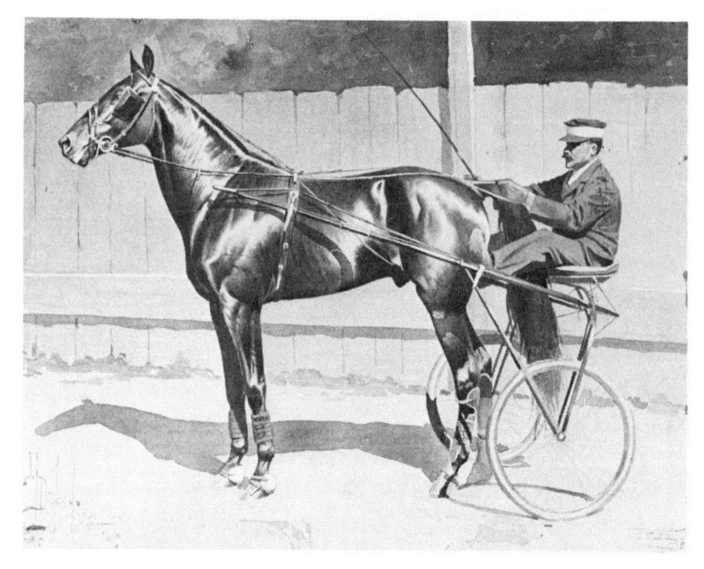

Bingen

Greyhound (Sep Palin)

and Mary Putney, a sterling race mare who died at North Randall, Ohio. Lu Princeton, another great racing stallion by San Francisco, after siring a few foals in this country, was sold for export. Finally, his son Vansandt was sold for export and became the leading sire of Europe. Vansandt was a World War II casualty and was last seen drawing a heavily loaded cart, driven by a Russian soldier.

Belwin was the son of McKinney that came from the East. Undefeated as a four-year-old, this horse later stood at Calumet Farm, then a prominent trotting nursery. His greatest son was Bunter, who produced speed for the short time that he was at a top-ranking eastern farm. Like McKinney's owner, Bunter's owner wanted his horse at home and he faded into oblivion in Ohio.

It was bad luck that made the McKinney family one of lesser importance today. Probably its greatest claim to fame came from a great-granddaughter of McKinney, Elizabeth, who had a gray colt named Greyhound. Admiration and honors were heaped on Greyhound, whose world trotting record stood from 1938 to 1969. Even the Thoroughbred people honored this trotter one year. His mile in 1:55 1/4 is still the fastest ever trotted over the famous old Lexington trotting track. Probably the naming of Greyhound as "trotter of the century" by the Red Mile and the Hall of Fame of the Trotter came closest to expressing his greatness. Greyhound, who raced for trainer-driver Sep Palin, was a real Kentucky product; he was bred by Henry Knight at Almahurst Farm, near Lexington. The world-famous gelding trotted twenty-five miles in 2:00 or faster and set twenty-five world records. Greyhound's world trotting mark of 1:55 1/4 had stood for thirty-one years when Nevele Pride (now a Kentucky sire) trotted a mile in 1:54 4/5 at Indianapolis.

It has often been pointed out as one of the oddities of breeding history that of the four sires recognized as the greatest contributors to today's breed, two (Bingen and Peter the Great) were owned by J. Malcolm Forbes and the other two

(McKinney and Axworthy) by William Simpson. Forbes bought both of his as colts, while Simpson sought out McKinney and Axworthy after they were already noted sires.

McKinney's twenty-five seasons of stud service meant, since he was a popular and fertile sire, that he got a very large number of foals. John Hervey, in *The American Trotter*, estimates that his sons and daughters may well have numbered over 1,000—surely more than the progeny of any of the other four great sires except Peter the Great.

HAMBLETONIAN with his owner, William Ryzdyk

FLORA TEMPLE

George Wilkes

Dexter

Axtell

Dan Patch

The Red Mile in the days of the first grandstand

Racing at The Red Mile

From The Red Mile clubhouse

A Standardbred mare and her foal in a Bluegrass pasture

McMurtry's Floral Hall, now a museum of Standardbred memorabilia

6

PATRIARCHS OF PACING

WHILE IN THE CITIES of the East the demand was for runners and trotters, on the American frontier the pacer was also valued. Here owners and breeders freely mixed the bloodlines of Thoroughbreds and pacers, producing some worthy progeny. Even in this part of the country, however, the pacer was far from a fashionable animal; his stud fee was typically just a few dollars and rarely did he serve any but the commonest mares.

The first attempt to popularize the gait was made in 1875, when the "Pacing Championship of the United States" was held at Mystic Park in Boston. Fourteen horses were in the original entry but only four were actually in the contest. The winner was Sleepy George. Instead of starting a pacing boom this was a farcical affair and was hailed with hoots of derision.

The man who really started pacing on its way to popularity was Governor John Todd of Ohio. He happened to own a gray pacer called Sweetzer and through his influence was able to get pacing races included in the prominent race meetings. Shortly after this, about 1879, the Big Four made their appearance. Mattie Hunter, Sleepy Tom, Rowdy Boy, and Lucy put on such sterling contests that they drew huge crowds and created the highest interest. It was the Big Four that finally made pacing acceptable to the American public and to the breeders.

The early pacers came from a section composed of Indiana,

Ohio, Kentucky, and Tennessee; Kentucky played a primary role in the formation of a real breed of pacers. The Kentucky pacers came from horses found in Canada, a nation that produced both trotters and pacers. These northern horses were generally called "Canucks" and the speediest of them came from the French section along the St. Lawrence River in Quebec. No records of their breeding or racing had ever been kept, so the history of this strain is a subject of speculation. The native horse stock had been brought from France and England. To this had been added a liberal infusion of the blood of the mysterious breed of Narragansett pacers, the origin of which is unknown. Early historians said that the animal that started the breed had been found swimming in the ocean or that it was washed ashore and was found eating the rushes in the region about Narragansett Bay in Rhode Island. But such stories are often told about a breed of unknown origin. The truth is no one knows where the Narragansetts came from and, in spite of their great popularity, no one knows what became of them.

The Canadians scoffed at the idea that the Narragansetts had helped to improve their horses, for importation was illegal. Besides the Indians were at war and horses could not be brought through the Indian territory. But Canada has a long unguarded border and there were long stretches of time when the Indians were peaceful; also there were persons not averse to smuggling horses over the line.

The useful and speedy Canadian pacers were seen as desirable by American buyers, but the Canadians in the section from which the best horses came had two difficulties in dealing with prospective buyers. One was a difference in speech and the other was that many were illiterate and so no written records of pedigrees were available. To the French Canadians this was not important, for lineage was passed on by word of mouth. Working in a crude and experimental way they produced a humble type of horse with such merit that it is found in the ancestry of today's fine Standardbreds.

The section along the St. Lawrence River whence these Canadian pacers came has long and severe winters, and people there had a passion for racing their pacing animals over the snow and ice. On Sunday the dwellers from outlying farms converged on the parish church, generally a stone structure flanked by the house of the priest, a tavern or two, a store, and a few dwellings. They bundled the family up and came in with their sleighs. After services the horses did not all travel at the same speed on the way home, mostly perhaps because some families were larger than others. This meant that some sleighs had to be passed and this in turn led to impromptu races.

The mania for racing reached such a stage that contests were held immediately after mass and the practice began to endanger the people who were leaving the church. It was necessary to enact a law banning fast driving within a certain distance of the sacred edifice. Law or no law the races went on, for those of sporting blood merely moved over to the nearby river, which was frozen with a smooth surface. It was these early Sunday contests that drew attention to the speed of the Canadian pacers.

The first of these pacers from north of the border to found a line of pacers that became famous in the United States was a horse called Copperbottom. This roan stallion stood for service in Lexington, Kentucky, in 1816. The advertising listed him as a seven-year-old "imported by Captain Jowitt." He came from Bolton, Quebec, by way of Michigan. Captain Perrin, one of his earlier owners, had offered to wager that Copperbottom could draw a cart carrying two men a distance of sixteen and a half miles in an hour. This horse did not found a great family, but his progeny were distributed over Indiana, Kentucky, Ohio, Tennessee, and Missouri. His sons produced quite a few pacers but the Copperbottoms were important mainly for their use in the breeding of saddle stock.

The first Canadian pacer to found an important family in the United States was Tom Hal. It is hard to trace this group because most of his descendants carried the same name—Tom Hal. Generally the owner's name was added to help identify the horse. But the difficulty was that when the horse changed owners he was likely to change names as well. It was possible that what appeared to be three different horses could be, in fact, only one animal that had had three owners.

Tom Hal was taken to Kentucky about 1824. John Wallace, the founder of the American Trotting Register, remarked of Tom Hal that "as was the custom in those days, he was called a Canadian, like all other pacing horses." The traditional story was that a Doctor Boswell got Tom Hal in Philadelphia and rode him all the way to Lexington. In 1828, while the horse was standing in Lexington in the charge of William L. Breckinridge, a Presbyterian minister, the advertising said that Tom Hal was brought into Kentucky by John T. Mason. This last is the version that has become accepted. Tom was himself a fast pacer and for some time he was in the hands of a Captain West of Georgetown. Later he went to Benjamin N. Shropshire of Harrison County and he died the property of this man. Like Copperbottom, Tom Hal was a roan in color and both families carried a liberal sprinkling of white in their coats.

The original Tom Hal was variously called Boswell's, Mason's, West's, and Shropshire's Tom Hal. It was in 1850 that Major M. B. Kittrell left Taylorsville, Tennessee, and went to Bourbon County, Kentucky, where he bought one of the many Tom Hals from Simon Kirtly to stand in Tennessee. This horse was represented as being sired by the original Tom Hal but the difference in ages suggests he was more likely by a son of the old horse, one called Bald Stockings who was also known as Lail's Tom Hal. The new purchase was named Kittrell's Tom Hal, and he died near Pulaski, Tennessee, after being seized by Federal troops during the Civil War.

In Tennessee this Kentucky stallion started the most famous branch of the family. Kittrell's Tom Hal had a son who, with true lack of originality, was named Gibson's Tom Hal and was sometimes called Tom Hal Jr. The latter was never trained for speed and was used as an all-purpose horse, more at home pulling a plow or a log wagon than racing. He was also used under saddle and generally brought in the cows at milking time. When the owner was in a sportive mood the stallion could be found in the group following the hounds at a fox hunt.

In 1874 Gibson's Tom Hal was owned by O. N. Fry of Mooresville, Tennessee, and was standing for a mere five-dollar service fee. But few mare owners were attracted to him because at that time mules were in greater demand. A neighbor of Fry had an attractive-looking mare but he refused to take her to Fry's horse; he was sending her to a jack because he could get fifty dollars for her mule colt. Fry guaranteed to buy the foal by his stallion for fifty dollars and the mare was sent to the pacer. When it came time to turn the yearling over to Fry, he found the little fellow half-starved and lousy to boot. Fry at first refused to accept this bedraggled animal but finally paid off because of his sympathy for the mare's owner, who was in dire need of provisions for his family. During the following year Fry made arrangements to sell the two-year-old to a tenant on his farm. The colt did all sorts of menial chores and the tenant family even rode him to church with the mother in the saddle and the kids in front and in back of her. The eldest son used the colt for courting and often let him stand out in even the most inclement weather most of the night, for the young man had things on his mind other than the care of the colt.

The wife of the tenant farmer fell ill and finally the doctor refused to make any more calls until the bill of six dollars was paid. He also demanded that Fry guarantee the payments for any future calls. In an attempt to defray this cost the farmer turned the colt back to Fry. Although in a pitiable condition from the hard usage, the colt responded to good care and eventually was ridden in some colt shows. The speed shown there prompted Fry to send the colt to a professional trainer. The rest is history, for the colt in question was Little Brown Jug and driver W. H. "Knapsack" McCarthy made this horse the first of the Hal family to become a world champion pacer.

After raising a couple of mules, the neighbor's mare was again taken to Gibson's Tom Hal and the result was Brown

Good Time (Frank Ervin)

Hal, who also became a champion pacing stallion. It is the only case in which two champions were half-brothers to a bunch of mules. It was Brown Hal who sired Star Pointer, the first harness horse in history to negotiate the mile distance in two minutes, a thirty-mile-an-hour clip. That was in 1897. There were many great horses among the Tennessee Hals, and although they failed to found a dominant male line, they did augment other bloodlines. One example of this is the immortal Adios, who carries a trace of the Hal blood. Another is the pony-sized Good Time, a great racehorse, whose sons and daughters have gone out from Castleton Farm in Kentucky each year to race against the best pacers in the nation.

One pacing family well represented today comes from Direct, a trotting-bred horse. This line also is deeply anchored in the soil of Kentucky. Richard West bred trotters at his Edge Hill Farm in Scott County, near Georgetown. When he sold his stallion Almont to General Withers Edge Hill was left without a sire. John W. Conley, a close friend and associate in many horse deals, was then managing the Spring Hill Farm of Harrison Durkee on Long Island. A son of Hambletonian called Dictator stood there; he was a brother to the world champion Dexter, whose bold trotting action made him the model for the weathervanes popular in that period.

It would be expected that a champion's full brother would be popular, but Dictator was neglected. This may have been because owner Durkee was so eccentric that he was unpopular with the horsemen. Conley was a shrewd man where horses were concerned and he induced West to lease Dictator and bring him to Georgetown. But the Kentucky breeders did not appreciate this stallion and after the second year Dictator was sent back to Long Island. As in the case of Abdallah, the Kentucky breeders had overlooked a great sire, for Dictator's foals were sensational. Dictator was later returned to Kentucky when he was purchased by H. C. MacDowell and brought to Ashland, in Lexington, where he died in 1893.

In 1879 Conley visited Kentucky to look over the foals of Dictator, the oldest of which were then two-year-olds. He wired West of his impending visit and upon his arrival there was a carriage waiting at the railway station, occupied by J. I. Case, H. D. McKinney, and West himself. They had driven only a short distance when a second carriage drew up behind them. The driver was George Brasfield, then superintendent of Edge Hill, and he happened to be driving a colt by Dictator.

Conley changed carriages and asked if the colt could trot. "A little," was the laconic reply. After pulling around the other vehicle the colt went away with a burst of speed that was surprising considering he was hauling two men in the carriage. This made Conley want the youngster and he found that he was riding with the owner, for Brasfield had purchased this colt from West when it was a yearling. Before nightfall the horse had changed ownership for $2,500, then a big price for an untried two-year-old. Immediately Conley sold a half interest to C. H. McConnell of Chicago. This buggy-pulling colt was Director, who was to become one of Kentucky's greatest early trotting racehorses.

Another son of Dictator was for sale at a price of $500 but Conley finally rejected the colt because he thought it was too small and pretty to be a good horse. One of the other members of the party, J. I. Case, purchased that one and named him Jay-Eye-See, after his own initials. This colt, which was supposed to be too small, was the first trotter to cover the mile distance in two minutes and ten seconds and also the first to go that fast on both the trot and the pace.

Conley left Director with Dr. Levi Herr at Lexington and the horse won five races and paid for himself before being sold for $10,000 to go to the West Coast. Monroe Salisbury, "the kingmaker," was then in the early stages of a remarkable career and it was he who took the young Director to his farm at Pleasanton, California.

Heavy shoes had been used on the colt and so the preparation for the racing season had to be careful to keep Director from going lame. After a late start he won four straight races. He came to his form fast the following year and embarked on a sparkling career, winning twelve of the fifteen races in which he started. He beat the best stake horses with but one exception—he couldn't handle his former paddock-mate Jay-Eye-See, the little colt who was rejected because he was too pretty.

Later, when placed in stud service, Director in 1885 produced a small son named Direct who had the unique distinction of being the only champion whose parents had raced against each other before producing him. Salisbury had given Director's service to a friend and neighbor who owned the good race mare Echora, and Direct was the result. Salisbury watched the little fellow and developed a great liking for him. When the neighbor came to get the mare and colt Salisbury asked how much money he would take for the colt and Direct changed hands for $1,000.

Monroe Salisbury had a money-winning stable but he was brutal to his horses. Many were raced so hard that they went lame and then were raced in that condition until they were ruined. It was Direct's misfortune to belong to a man whose main concern was to race for personal profit. Outside of this, Salisbury always went first class and he would come over the mountains each spring to race the crack eastern horses for the big purses. His stable was always fully equipped, even to a cook tent so that the men could be properly fed. Salisbury was more concerned about the welfare of his grooms than he was about his horses.

One morning Salisbury left the hotel earlier than usual in order to witness the training of one of his horses. By the time he arrived at the track the grooms had been fed, so Salisbury headed for the cook tent for his breakfast. He ordered ham and eggs. The cook sadly shook his head and said that they were all gone. The owner stomped out angrily and soon built up so much steam that he decided to go back to the tent and give the cook a severe lecture concerning the stocking of ample supplies. As he stepped through the tent flap he stopped aghast, for there sat the big cook contentedly devouring a huge platter of ham and eggs.

Although Direct heads a prominent family of pacers he was not a natural pacer himself. He began his racing career

ADIOS (Frank Ervin)

BRET HANOVER
(Frank Ervin)

as a trotter but went lame and developed a tendency to amble. When the decision was made to convert him to the pace it was necessary at first to use hobbles, but once he developed the pacing gait these corrective straps to make the legs on each side move in unison were not required.

Probably no champion was more brutally misused than Direct. He not only raced against but beat the best pacers of his day. He was so lame that even the spectators in the grandstand often expressed pity for him. At Columbus, Ohio, during an extended contest, Direct twice threw himself in the stall between heats and lay on his back in order to ease the pain in his legs.

In spite of the piteous shape he was in Direct was required to make a campaign of twenty-one races. In some of these it was almost impossible to get him to the track. While in this condition he set the world record for pacers. Like his sire, Director, that great trotter from Kentucky, Direct was finally ruined for racing and was retired to the stud.

His first full season was in 1893 and from this came a son called Directum Kelley who, before he was exported, had a son named Directum I, one of the most beautiful and fastest of our champions. Another son of Direct, sired before his retirement, was the two-year-old champion Directly, whose performances in 1894 gave rise to a demand for Direct's services in Tennessee. Salisbury allowed him to stand in that state for one year, 1895.

This single season was a banner year, for it started one of the greatest pacing families of today. From Direct came the line of speedy pacers and outstanding sires that includes Direct Hal, Walter Direct, Napoleon Direct, and Billy Direct, a champion for many years. The present representative of this family is Tar Heel, one of the greatest living sires of extreme speed. (Tar Heel stands at Hanover Shoe Farm, in Pennsylvania, perhaps the greatest rival of modern Kentucky nurseries.)

Richard West's Edge Hill Farm only lasted about one decade but it exerted a great influence and was a prominent factor in the founding of both the trotting and pacing families. No man connected with the trotting industry was ever more respected and beloved by all who came in contact with him.

Kentucky also had a hand in the formation of another great line of pacers. One of the victims of the Civil War raid on Woodburn was Bay Chief, a young stallion with only a few foals. The farms did not wish to expend time and money on these few, by a practically unknown sire, so they got rid of them as soon as possible. One of the Bay Chief colts went to Wisconsin and was known as Stevens' Bald Chief. His daughter Minnehaha was sent to California and became the granddam of many good horses. One of her grandsons was Chimes, who came from Palo Alto Farm, located on the site where Stanford University now stands.

Chimes came east to race and took his record in the Kentucky three-year-old trot at Lexington. While it was in operation this was the oldest fixture for that age in the trotting world. Later Chimes went to the Village Farm where he was crossed on another line from Kentucky, that of Mambrino King. This line produced the trotting champion The Abbot and also a full brother, The Abbe, who turned to pacing. The line has given the sport some of its greatest speed as well as outstanding broodmares and sires. The best known are Adios, the great sire of pacers, Good Time and Bret Hanover at Castleton Farm, and Stoner Creek Stud's Meadow Skipper. Another line from The Abbe, the descendants of Bert Abbe, has helped to improve the breed.

The cross producing the greatest pacing speed today is the blood of Direct and The Abbe, both of whom owe a great deal to the state of Kentucky.

7

BELGRAVIAN DAMES

BELGRAVIA WAS A FASHIONABLE DISTRICT located in the West End of London. Here the aristocratic society lived and the women were called Belgravian Dames. Harness horses too have their Belgravian Dames. These are the matriarchs whose blood has bred on and on through their daughters.

The covering of a mile in two minutes or less was at first a rarity. But as tracks were better constructed and modern equipment appeared, and as the breed was improved by selective breeding and crossing of bloodlines, harness horses have consistently increased in speed. The result has been an increasing number of these fast miles each year. In 1943, during the war years, only three of the fast miles were recorded, but in 1976 a new record was set with 1,847 of these miles of super speed.

It has been only a little over a quarter of a century since the students of bloodlines noticed that the family trees of the super speedsters invariably led back to certain matrons. These are called foundation mares. The Belgravian Dames of this select group are those who have more than sixty descendants with records of two minutes or better.

One fact stands out clearly about the families to be discussed: ten out of eleven came from Kentucky. In most of these pedigrees the blood of Mambrino Chief can be found. This is the same line that gave Kentucky its initial start on the long road to greatness in the harness sport. Many of these same foundation mares have a cross to George Wilkes somewhere along the line. The early efforts of the Kentucky breeders have really paid off.

The top ten of the great Kentucky matrons are discussed below in the order of their standing at the end of 1976. For those interested in statistics, the number of two-minute performers to her credit at that time is given after the name of each mare.

MINNEHAHA (129). This family started in Kentucky, moved to California, and finally came home again to the Bluegrass.

When the Woodburn raid was reported it was said that Abdallah and "another horse" were killed. The unidentified animal was Bay Chief. The guerrilla chief rode him away, so that the horse became a prime target in the running battle that followed the raid. Shot through the muzzle, flanks, and one leg, the stallion gamely continued for two miles before falling. All attempts to save him were futile. As there were only a few colts by Bay Chief at Woodburn the farm was anxious to dispose of the offspring of this young and untried sire. One of these was Bald Chief, who was purchased by George C. Stevens, at that time one of the largest breeders in Wisconsin. His farm was located at West Allis, on the site where the Wisconsin State Fair is now held. In the vogue of the times the new owner added his name to his purchase—Stevens' Bald Chief. The highlight of the stallion's career was reached as soon as he arrived at his new home, for that was when he sired Minnehaha.

She went, in a group of six horses, to L. J. Rose's new nursery in California. Rose had gone west in the wake of the forty-niners and fought Indians and wild beasts with many narrow escapes, but through all of this he managed to keep his band of horses with him. In a score of years Rose amassed a fortune with his vineyards, fine wines, and a famous brandy. It was Rose who named the mare Minnehaha, laughing water—rather appropriate for a maker of wines and brandy.

When Rose decided to raise harness horses he had his citrus orchards and a section of the vineyards cleared. Here he built barns, paddocks, and even a track, three-quarters of a mile in circumference. He then headed for Kentucky to buy stock. The transcontinental railroad had just been completed and this consignment of horses was one of the first to go to the West Coast by rail. In the group with Minnehaha was a young stallion called The Moor.

None of the men employed by Rose knew anything about training horses and their crude methods soon ruined Minnehaha for racing purposes. She was sent to the breeding section of the farm and during her life was mated only with The Moor and his sons and grandsons.

Her first foal was Beautiful Bells, the queen of Leland Stanford's Palo Alto Farm. Later the mare was to be known as "the empress of broodmares" and was famous for her champion offspring. Her sixth foal was Eva, a mare who was crossed with Guy Wilkes, another Kentucky product that had been taken to the West Coast. From these unions came three sisters that some pedigree men consider the head of this family. These were the Thompson sisters, Madam, Lydia, and Tillie.

Patchen Wilkes Farm obtained the three sisters and while in Lexington they and Peter the Great grew famous together. One of this line was Miss Stokes, a champion yearling, and from her came Tillie Brooke, the first trotting mare to race in two minutes. Another of the line was Miss Pierette, a great matron who was the darling of two stock farms. Minnehaha made a fortune for Rose and is the matriarch of a family that has moved up rapidly. In this group are many good pacing horses as well as trotters.

MEDIO (128). The early history of this family is practically unknown. Medio's sire was Cooper Medium, a horse of mystery. Although he had been registered with the American Trotting Register, nothing else was known about him and many thought that he had died young. There was no man who knew more about Kentucky horsemen and horses than the late Jesse Shuff, a turf writer for the *Lexington Herald* and national racing magazines, and he undertook the task of digging out the background of Cooper Medium.

Kentucky has a reputation for being a place where family history is important, and Jesse found that people not only knew about their relatives but also knew about the horses that had been owned by their relatives and friends. Cooper Medium had come from Georgetown, right in the section where Jesse had grown up. The breeder was John Haggin Cooper, who inherited his love of horses from both the Coopers and the Haggins. They had bred most of the horses in Cooper Medium's pedigree, which went back to Mag Cooper, a mare with a mind of her own. During the Civil War, both armies coveted the Kentucky horses for their cavalry units, and the Federal troops decided that they wanted Mag Cooper. There was a rough battle and after three stormy sessions in the breaking pits it appeared that Mag leaned toward the Confederacy. The Federal troopers decided that they had had enough of her and raised the white flag.

This unruly temper apparently carried on to Cooper Medium, for in unearthing his history Jesse learned that the stallion got out of doing things he didn't care for by kicking. This caused him to be exiled to Texas and there he passed into obscurity. But early in his career the reprobate had sired Medio, the matriarch of this group. It was probably the only good thing Cooper Medium ever did.

Medio was brought to Lexington from Muir, Kentucky, by Major F. P. Johnson, who had first raised saddle horses and gone on to Thoroughbreds and finally to harness horses. He bought Medio with the intention of using his stallion King Clay for her consort. But at this time D. C. Parmenter of Berlin, Wisconsin, was in the process of establishing his Riverside Park and, like so many of the northern breeders, he turned to Kentucky for his basic stock. He not only bought Medio but also her daughter Marble. The latter carried on the tradition of Mag Cooper and her repertoire lacked little in the line of unruliness. In disgust, Parmenter once offered her for sale at a bargain price but could find no one interested. So the unruly one was taken back to the farm on the banks of the Fox River. She was then initiated into the farm harem and became a great broodmare.

One of Marble's daughters was named Bertha C., after the daughter of her trainer, J. B. Chandler. This one was a true daughter of her mother, though not quite as unruly on the track. When she wanted to trot she had blinding speed, but more often she would put on a mediocre performance. Bertha C. always carried her ears pinned back, a habit which became her trademark. She allowed no stranger near her and even her driver had to carry a whip when he went into her stall.

After a virulent form of influenza hit Parmenter's stable, none of the horses were able to get back to their racing form and he decided to retire from racing. He sold his farm and all of his horses. Bertha C. (later registered as Miss Bertha C.) went to A. B. Coxe, who had been a great athlete at Yale and was in the process of establishing his Nawbeek Farm at Paoli, Pennsylvania. At about the same time, Coxe went to Lexington in search of a colt that would not only be suitable for racing in the big colt stakes but who also could be used as a sire after his racing days were over. R. L. Nash was anxious to rid himself of a colt that had been given to him for settling an estate, so he sold Dillon Axworthy to Coxe. While the great colt was being prepared for his three-year-old racing season he was also bred to three mares to test his fertility. One of these was Bertha C. and her daughter was Miss Bertha Dillon, who was to be a champion four-year-old mare.

After the death of A. B. Coxe, in 1926, his entire horse holdings were put up for sale. L. B. Sheppard of Hanover, Pennsylvania, purchased all of the stallions, mares, and foals and moved them to Hanover Shoe Farm. This was the start of one of the largest and most prominent harness horse nurseries in the world. Here Miss Bertha Dillon was the first to have three daughters with two-minute speed. The mare was sent to one of the famous Kentucky stallions and died in Lexington. She is buried in the small plot at the Lexington trotting track, where Red Mile patrons sitting in the grandstand can see the headstone at the head of the homestretch.

The Medios have produced more fast trotters than any of the other families, but they also have great pacers. Bret Hanover, one of the Medios, won the pacing crown when he paced the Red Mile in 1:53 3/5 for trainer Frank Ervin.

JESSIE PEPPER (126). The biblical passage concerning a stone rejected by the builders that later became the keystone of the building could almost fit Jessie Pepper. Alma Mater was the star of Dr. Talbert's Inwood Farm and Jessie Pepper stood in the shadow of her paddock-mate.

Jessie was sired by old Mambrino Chief himself and was one of the last foals of Kentucky's first great trotting sire. It was an oddity that Mambrino Chief's two greatest daughters had eye trouble. Kentucky's first great race mare was Lady Thorn, one of the first foals of Mambrino Chief, and she had the sight of only one eye. One of his last foals was his greatest daughter and she was entirely blind. Jessie Pepper never saw any of her foals.

As her name indicates, the mare was bred by Colonel R. P. Pepper at his South Elkhorn Stud. Jessie went to Dr. Talbert for $300, so that this man for a total of $500 purchased two of the greatest broodmares in the history of harness horses. Jessie Pepper had eighteen foals, the last at the extreme age of twenty-eight, just one year before her death. The great matron was a plain-looking mare, like most of the Mambrino Chiefs, but she did not pass this on; her foals were attractive.

Seven was a lucky number in the case of Jessie, for her seventh was the foal named Annabel. This daughter of George Wilkes has a line of descent that carries some of the greatest names in trotdom. The first of the family to go in two minutes was Rose Scott, a mare foaled at Poplar Hill Farm near Lexington, who is the third dam of Tar Heel, a champion on the track and a leading sire of extreme speed. A full brother to Rose was Scotland, one of the greatest of trotting sires. The fastest of the group was Rosalind 1:56 3/4, a mare that wore the crown for trotting queens from 1938 until a new queen was crowned in 1974.

When she passed on, Jessie Pepper was the last living daughter of Mambrino Chief and there were few at that time who realized what a great contribution this mare was to make to the sport. Her fame was posthumous and the rejected stone became the keystone for the foundation mares. When study of maternal families began it was Jessie Pepper who was first noticed and she consistently led all mares until 1974.

KATHLEEN (98). The foundation mare with the longest family tree is Kathleen. Her Thoroughbred antecedents can be traced back to about 1700 in England. One of these, Little Miss, was in the Thoroughbred section of Woodburn. Only once in her life was she mated with a trotter, and the foal was Kathleen, one of the last foals of Pilot Jr. The mare was sold to Colonel R. P. Pepper and Major H. C. McDowell. The latter soon bought out his partner. At first McDowell was located at Woodlake Farm, about seven miles from Frankfort, Kentucky, but in 1882 he purchased Ashland. He may have been influenced in this by his wife, who was a granddaughter of Henry Clay. The tenure of McDowell at Ashland explains the trotting registers and the yearbooks with the racing results, as well as an oil painting of the trotting mare Trinket, which can be found today in the former home of the great statesman. Kathleen and her daughter Ethelwyn were among the horses moved to Ashland.

McDowell was public-spirited as well as a shrewd horseman. He gave Fayette County the speedway, a strip of ground starting at the Ashland entrance and extending about one and a half miles. The course had been designed for use in training horses.

McDowell had many of the noted broodmares of his day while he was at Ashland and from these came champions and stakes winners. It wasn't long before this nursery ranked among the best of breeding establishments, not only in the state but also in the nation. One of Ashland's products was Extasy, who won the two-year-old pacing division of the Kentucky Futurity, setting a world record for that age and gait that was not beaten until 1927, twenty-nine years later. The following year Extasy was back, this time as a trotter, and she won a heat in the senior trotting division of this oldest of futurities.

From the line of Extasy came Nedda, who held the trotting

crown for mares for nineteen years. Further on in the line is the pint-sized pacer Good Time, among the best sires of extreme speed. This former occupant of the Castleton stud barn has a pedigree going back to 1700 on both sides of the family tree. One daughter of Extasy was given to David M. Look and from Petrex he got his great sire, Spencer. This great colt trotter had a part in another of the great Standardbred crosses started in Kentucky: the blood of Spencer, who stood at Castleton, and the blood of Scotland, from nearby Walnut Hall.

From the line of Ethelwyn also came McLin. Purchased the week before the Hambletonian Stake, which he won under the name of McLin Hanover, he later equaled the three-year-old race record of Protector before he was exported to Europe.

Starting at Woodlake and reaching maturity at Ashland, this Lexington family has been responsible for some of the greatest horses in the harness sport.

MAMIE (84). Philemon P. Parrish had served in the Army of Northern Virginia under General Robert E. Lee and had meager finances when he returned to Midway, Kentucky, after the conflict. He gamely spent $200 for a young mare called Kit for use as a driving horse and for general farm work. The mare was leased to a friend for a year to raise a foal and after that Parrish decided that he would raise a few horses himself.

Kit was sent to a nearby stallion called Star Almont, owned by Patrick Nolan. With a name like that a person would expect to find a liberal attitude toward wagering and racing, as well as a different religion, but Nolan was an elder in the Presbyterian church and he raced no horses. He had bought Star Almont solely to raise fine harness stock for use over the roads.

Kit's filly by Star Almont was Mamie and she was to be the only foal that Kit would produce for Parrish. As a young mare Mamie was used by the owner's son to ride to school. Young Parrish was much the same as the present vintage of youngsters, for he did a little drag racing against a boy on a pony. Trainer Mike Bowerman happened to see one of these impromptu contests and noted that Mamie held true to the trot while the pony was galloping. Mike looked up Parrish and persuaded him to have Mamie trained as a racing prospect, but lameness caused her early retirement to matronly duties. She had six foals and all were females. It is remarkable that she was able to start a lasting family, as three of Mamie's girls were exported.

Many members of the Mamie family carried the name Leyburn. It seems that owner Parrish had read the novel *Robert Elsmere*, in which there are characters named Catherine and Rose Leyburn, and he used these names for Mamie's foals. As he did well with the name he continued to use it. John Madden took a liking to the Leyburns and he delighted in telling a certain story whenever Parrish was present. He said that Parrish had put a price of ten thousand dollars on one of his horses and Madden had countered with an offer of half of that amount. Parrish had promptly accepted with the comment that he never let a small difference in price prevent him from closing a horse deal. The Leyburns were Madden's favorite group and most of those he had he obtained in one block from Parrish.

The best line came from Rose Leyburn and one of these was Margaret Parrish. Her second foal was Arion Guy, the first colt to trot a mile in two minutes. Sixteen years later she had Margaret Castleton, the first trotter to go in two minutes while driven by a woman, Mrs. H. Willis Nichols, Jr., of Walnut Hall Farm. It was at Walnut Hall that Margaret Parrish and her daughters attained their lasting fame. One representative of this family is Protector. While winning the Kentucky Futurity he set a race record for three-year-old trotters that stood for thirty years.

Up until the last few years the family of Mamie led in the production of fast trotters and many of the noted sires and matrons come from the Mamie family. It has now been supplanted by the Medio family as far as trotters are concerned. The foreign buyers liked Mamie and, as early as 1882, Wilbur N. beat all the best trotters in Europe. Here in the United

States one of the family was Demon Hanover, a great racehorse himself and sensational in his early stud performance. Unfortunately, he had an early death, just a couple of years after he had been syndicated for half a million dollars. Money is also connected with the Mamies, for Fresh Yankee was the first American horse to bank over a million dollars from her race earnings. In the line of speed, Albatross recorded the most miles in two minutes or better in the history of the sport.

MIDNIGHT (79). (This is also known as the Emily Ellen family.) Like so many of the foundation lines, this one started at Woodburn Farm. This daughter of Pilot Jr. was later sold to Colonel Richard West and while his property she became the only foundation mare to herself produce a champion trotter, Jay-Eye-See. Before she left Woodburn, Midnight had a daughter named Noontide. A daughter of the latter was Rosy Morn and she was one of the selections made by a group who revived trotting at The Hermitage, former Nashville, Tennessee, home of Andrew Jackson. When David M. Look was starting at Castleton one of the first mares he selected was Morning Bells, a daughter of Rosy Morn.

His father, Samuel J. Look, of Louisville, suggested that the mare be mated with Todd, Bingen's finest son. The elder Look never lived to see the result of his suggestion as he died on the same day that Emily Ellen was born. David M. Look was a super salesman but he never sold Emily and he also kept most of her produce. Emily Ellen was the base on which Castleton Farm built its present fame. She had four sons who produced two-minute speed.

Emily was blind for a number of years before she died at the extremely advanced age of thirty-eight. The achievements of Emily Ellen overshadowed Midnight's so much that some of today's pedigree experts call this the Emily Ellen family. But Midnight had another branch that came to full bloom later.

Lady Kerner went to Stony Ford Farm, near Goshen, New York, and became one of their prized mares. Kentucky Prince had gone to this farm earlier and it was his son who was used on the line of Midnight to sire Polly Pry. At this time Billy

Dickerson had charge of Joe Patchen, for the sire of the great Dan Patch was then standing at Goshen. Mrs. Dickerson, now over ninety, remembers riding behind Joe Patchen as her husband exercised the pacer over the streets of the little village. The daughter of Joe and Polly was Fan Patch, winner of the historic Walnut Hall Cup at Lexington in 1913.

A daughter of Fan, Great Patch, was acquired by Henry Knight when he started his large breeding operations at Almahurst. The first foal from this new venture was the great race mare Little Lie whose line is credited with Elma, a star race mare on two continents. There was also a juvenile delinquent named Ayres, who reformed and became a Hambletonian winner; he is still a champion colt over a half-mile track. Noble Victory is another to hold the stallion record, but Emily's branch also had a champion stallion in Spencer Scott.

MAMBRINO BEAUTY (77). (This is also known as the Nervolo Belle family.) In 1892 Grant Lee Knight bought a filly for $150. A brother, W. P. Knight, who generally got what he wanted, handed over the same amount. Although the mare had not been purchased for resale, he took over the ownership. Called Josephine Knight, the mare never raced, for an attack of pinkeye left her permanently blind. Knight's brother-in-law, Scott Hudson, was training a great stable of horses at the Lexington track, so Knight sent his blind mare to be bred to Jay McGregor, the star of the Hudson stable. Hudson was loathe to use his best horse, who was in his final preparations for the races. Instead he used the pacer Nervolo. This mismating caused hard feelings in the family for some years, but it was one of the most fortunate mistakes in harness horse history, for the result was Nervolo Belle.

Like her dam, she too was seriously ill but Miss Carrie Knight nursed her back to health with a diet consisting mainly of raw eggs and bourbon. When W. P. Knight died the ownership of the mare passed to his daughter, who had saved the filly's life, but G. L. Knight did all of the managing. After being used on the roads, Nervolo Belle began her career as a broodmare and was sent to Patchen Wilkes Farm to be bred to Peter the Great. W. E. D. Stokes did not consider the pedigree of this mare sufficiently attractive for his famous sire and he used another stallion instead. The following season there was a definite agreement and Nervolo Belle was bred to Peter the Great. The foal was Peter Volo, the only trotter to hold the record for the ages of one, two, three, and four. According to his agreement with Knight, Stokes exercised his option and picked up the colt at weaning time. Later he took full credit for Peter Volo, the son of the mare he had once spurned, but this was after the horse had started to chalk up championships. In his extensive racing career Peter Volo was beaten only once. Nervolo Belle's next foal was Volga and she was undefeated. She was a full sister to Peter Volo.

Later Nervolo Belle sold for $10,000 and went to Laurel Hall Farm at Indianapolis. While in Indiana she went through two dispersal sales and finally came back to Lexington to Calumet Farm. G. L. Knight was a man of considerable sentiment and when he died it was found that he had made final arrangements for Nervolo Belle, who was then a pensioner at Calumet Farm, for she had not been sold in the dispersal of the Calumet trotters. Nervolo Belle was to be brought back to the Knight farm so that she might die on the same place where she had been foaled. Today the great matron sleeps in the Almahurst burial plot.

The largest branch of the family comes from Volga, now known as Volga E., the initial having been added in honor of one of the mare's owners. In this group is the noted trotting queen Rosalind 1:56 3/4, who held her crown for thirty-six years, the longest reign of any mare. Rosalind (who was dethroned in 1974 by Colonial Charm, another Lexington product) was a product of thirty years of Ben White's wizardry in training. He raced most of the horses in her pedigree, including Volga.

The most prominent sires to come from this group were Peter Volo and his son Volomite, the first to "production line" extreme speed while in the stud. People working with pedigrees of good horses soon note the frequency with which the

Rosalind (Ben White)

Colonial Charm (Garland Garnsey)

name of Nervolo Belle appears. A majority of the best harness horses carry this strain.

MAID OF HONOR (68). (This is also known as the Lizzie Witherspoon or Isotta family.) When C. J. Hamlin went to Kentucky to buy Mambrino King, in 1882, he also bought the mare Betty Mac, who came from a line started by Lister Witherspoon of Midway, Kentucky, from an unknown mare he had purchased. The smart thing that Witherspoon did was to pass by the cheap stallions. He bred only to the best. One of this line was Lizzie Witherspoon, a daughter of Almont, who took her best record at the age of ten. Lizzie was sent to Red Wilkes, a son of George Wilkes, to get Betty Mac, the mare purchased by Hamlin. When Hamlin crossed his two purchases he got Maid of Honor.

In Lexington, the Reverend T. C. Stackhouse had a two-year-old, Rex Americus, who could show some speed. Hamlin gave $15,000 for Rex Americus, a horse that did not live up to expectations but was part of one of the greatest hoaxes of the harness sport. Hamlin sent out stories concerning the great speed of Rex Americus and this publicity was so effective that other owners were afraid to start against the stallion.

His daughter out of Maid of Honor was American Belle. She went to Nawbeek Farm, where A. B. Coxe had become one of the most successful breeders in the sport. American Belle was a shy breeder. She did not have many foals but her maternal instinct was strong; she loved to watch over the little foals. Coxe said that once, when another mare died, the barren American Belle adopted the orphan foal and in a few days made milk and nursed it.

One of the few foals of American Belle was The Zombro Belle and the most famous offspring of the latter was Isotta, who was heavy in foal when A. B. Coxe died of diabetes in 1926. The foal was Zombro Hanover, the first of the family to race in two minutes.

Many great horses have come from this family and many of this line have been exported. The star of the troop was registered as Thankful's Major and he is the fastest trotter in history. Better known as Nevele Pride, for his name was

VOLOMITE

NEVELE PRIDE (Stanley Dancer)

changed, he is only the second stallion to hold the best trotting record and also has the fastest mile over a half-mile track. Nevele Pride stands at Stoner Creek Stud near Paris, Kentucky.

MAGGIE H. (65). Thornton Moore, of Lexington, took a Thoroughbred mare, Sally Sovereign, to Sentinel and the daughter foaled was Lady Sentinel, the dam of Maggie H. Few horse deals took place around Lexington at this time with which Mike Bowerman and his brother George were not connected in some way. They bought Maggie and had a heavy influence on her line. In 1891 the brothers established a farm which they called Wilton's Place. Later it was sold to W. E. D. Stokes and he renamed it Patchen Wilkes Farm.

Maggie H. spent most of her life in the Bluegrass and was once owned by W. C. France, owner of the Highlawn Farm on Old Frankfort Pike. Her daughter was Maggie Wilton, a champion in her day, but when she was bought by Patchen Wilkes Farm her name was changed to The Widow. This mare is noted for her sons, one of which sired the dam of the great Canadian mare, Widow Grattan. In one of Widow Grattan's races there was some close riding in the first turn and her driver was tipped out of the sulky. The little mare went on and turned in a perfect performance and came on in the final strides to finish in front. Her driver never saw anything funny in the frequent remark that this mare didn't need a driver.

One of this line was Widow Maggie, a favorite of W. H. L. McCourtie who was then living in Minneapolis, Minnesota. He had gone to Lexington with a party of friends to enjoy the Trots and one of the group talked him into buying a yearling. This was Widow Maggie, who spent the rest of her life in the McCourtie ownership and gave her owner several champions. The best of them was a great colt and many-time champion, Mr. McElwyn. He started his stud career at the Lexington track and later became one of the top echelon of sires.

One of Maggie's daughters was responsible for the champion Lee Axworthy, who died prematurely at Castleton. This daughter was the first of the family to follow the two-minute trend. Maggie H. has had a hand in some way in the consecutive line of stallion champions ever since Lee Axworthy, who reigned from 1916 to 1941. Then came Spencer Scott and nine years later it was Star's Pride. Harry Pownall came back a decade later with Matastar, a son of Star's Pride, the only time that a driver has ever set the stallion record with both sire and son. There was a four-year gap before Noble Victory came along and in 1969 it was the current king, Nevele Pride. Maggie H. appears in the family trees of all of these horses.

ESTHER (65). Only one Thoroughbred ranks as a foundation mare with the harness horses. In 1879 Leland Stanford sent his superintendent, Harrison Covey, east to buy a carload of Thoroughbred mares for an experiment in crossing them with his sire Electioneer, whose foals were all trotters—with one exception. One man, tired of listening to Stanford's boasts that Electioneer had never sired a pacer, spent several years training Peruvian Bitters, a son of Electioneer, to pace. The horse spent his whole life in training before he finally managed to get a record on the pacing gait.

In Kentucky one of the mares selected by Covey for her good disposition and her natural trotting action was Esther. As a matron she had two daughters who helped the trotters. One was Expressive, a posthumous foal of Electioneer, who was one of the greatest of the three-year-old mares. She raced hard and long and generally was matched against seasoned aged campaigners.

When retired, this mare had foals that gave rise to the belief that she had left all of her vitality on the racing strip. Her later foals were better, however. She had a son, Atlantic Express, who sired the trotting queen Nedda. Another of his daughters came up with Dean Hanover, a champion three-year-old trotter who was driven to his record at Lexington by twelve-year-old Alma Sheppard.

The second daughter of Esther was Mendocita, who never raced, but her daughter gave the sport the outstanding sire Volomite. This branch of the family was badly hurt by the death of another daughter, Mary Putney, a brilliant race mare who died suddenly during the racing season.

The above mares all came from Kentucky. The only alien Belgravian Dame is MISS COPELAND (61), who ranks comparatively low in the list. This mare came from New York State. Her dam was a mare of unknown breeding with no name, so she is called the Copeland mare; she is believed to have been sired by a Morgan horse. It is the only northern family in this group and it became famous before any of its branches went to Kentucky. Fruity Worthy (fifth generation) went to Walnut Hall Farm and started a line of fast horses. Justissima (seventh generation) spent several years in Kentucky but made her greatest contributions in another state.

From this group came Titan Hanover, the first two-year-old to trot in two minutes, and sires such as Nibble Hanover, Knight Dream, and Sampson Direct. The family has great racehorses, sires, and top broodmares, and is the only family in this select group that Kentucky cannot claim. But ten to one is considered to be good odds.

It was John E. Madden, master of Hamburg Place and known as "the wizard of the turf," who said that the mare is the larger part of the foal and the stallion is the larger part of the nursery. (Madden is the only breeder ever to produce winners of both the Kentucky Derby and the Kentucky Futurity. He won the Derby with Sir Barton in 1919, and in that same year the Futurity was won by the Madden-bred Periscope.)

It was this recognition of the significance of the maternal influence that caused the harness horses to progress so steadily. The modern Standardbred is indebted to Kentucky for the serious planning that produced the outstanding horses of the Bluegrass breeders. The great harness horse families have taken root and blossomed in Kentucky. Their great matrons have continually produced outstanding horses and their speed breeds on and on. It is no wonder that the owners of other big farms are in attendance at the Kentucky sales. When a dispersal sale is held the competition is keen.

8

PRODUCERS OF CHAMPIONS

SOME OF THE GREAT early farms of Kentucky have already been discussed, from Woodburn, which affects all of today's harness horses, to Walnut Hall, which for nearly eighty-five years has influenced breeding and racing methods, especially in early colt speed and futurity winners. Some of the other famous old Standardbred nurseries in the Bluegrass are described in this chapter.

CALUMET. J. W. Bailey, a U.S. Senator from Texas, raised many good trotters at Fairland Farm—which is known today as Calumet—before selling it to U. G. Sanders about 1905. At about that time Henry Schlessinger of Milwaukee had the unbeaten Belwin as a sire at his Harvest Farm there. He sold all but a few of his best mares and moved his operations to R. S. Strader's Forkland Farm at Lexington. He then purchased Fairland from Sanders, who had owned it for only two years. Then, in 1924, another out-of-state buyer appeared: William M. Wright, of Chicago, bought Fairland and gave $50,000 for Belwin.

Wright was originally from Dayton, Ohio, and was a cousin of the famous Wright brothers, but his feet were solidly on the ground. Wright had worked hard, amassed a fortune, and one year paid the highest income tax of any Chicagoan. Having inherited a love of horses from his father, he went to nearby Libertyville, Illinois, and made frequent visits to the famous Grattan Farm of F. E. Marsh. A country home was purchased at Libertyville and here Wright bred a few horses. One was named after a close friend, Peter Manning, and held the trotting crown as long as Wright lived. Ill health delayed Wright's plans for expanding the breeding establishment. He underwent seven serious operations and the surgeons gave no hope for his recovery, but he survived. When he was seventy-five it was announced at the Trots that W. M. Wright had bought Fairland Farm. He immediately changed the name to Calumet Farm, in honor of a leading brand of baking powder that had helped to make his fortune.

The farm was reconditioned and only the best stallions and mares were added. The Calumet horses became very prominent and began winning most of the big races. The master of Calumet was interested from the start in the Hambletonian Stake and, although he lived in Kentucky, he did all in his power to make the Hambletonian the greatest three-year-old race for harness horses in the nation. His main ambition was to win this prestige race with one of his horses. The ambition was realized but he never knew it. In May, 1931, after an illness of several days, a stroke left Wright in a coma. He never knew that Calumet Butler did win the big race, just two weeks before his owner died at Calumet Farm.

The Calumet trotters were sold in groups at several dispersal sales. Buyers were eager to bid on the racing stock as well as on the breeding stock that had produced such great horses. The owners of the young Hanover Shoe Farm were active in the bidding and many of the Calumet horses went to Pennsylvania. From this basic stock, a generation later, came a colt that was to record a milestone in trotting history. When Harry Pownall drove Titan Hanover in two minutes at Lexington in 1944 it was the first time that a colt of that age, regardless of gait, ever went that fast. The Arden Homestead champion was far ahead of his time, for it was nine years before the two-year-old pacers achieved the two-minute mark. Titan's record was equaled ten years later, but it was thirteen years before it was beaten by a two-year-old trotter.

Although only in operation as a Standardbred nursery for seven years, Calumet Farm still exerts a heavy influence on the speed of today's trotters and pacers.

POPLAR HILL. Nancy Hanks, one of Kentucky's sensational trotting champions, was born in a little field near the springhouse on the farm of Hart Boswell, one of Kentucky's earliest and most respected breeders. Later this acreage was to become the Poplar Hill Farm on Russell Cave Pike. Henry Oliver of Pittsburgh, Pennsylvania, had a money winner in Peter Scott. He also had one mare, Roya McKinney, a daughter of the foundation sire McKinney. With just two horses he started breeding operations at Poplar Hill. No other person ever had such success with just a pair of horses, for he bred a great trio—Scotland, Rose Scott, and Highland Scott. All were great racehorses and Scotland became one of the great sires. Rose Scott, winner of the 1931 Kentucky Futurity, was a famous broodmare and is in many of the good pedigrees. With Highland Scott, Mrs. E. Roland Harriman earned a spot in the Standardbred annals when she became the first woman to drive a mile in two minutes. Unlike his brother and sister, Highland Scott was a pacer but he did not need hobbles to make him keep to his gait.

The farm later passed to Mr. and Mrs. Rex C. Larkin of Chicago, and operations were continued on a larger scale. The horses with the prefix Poplar in their names became better and better and had reached an important position in turfdom before the tragic death of the owners in an airplane crash near Cincinnati in 1965.

One of the Larkins' favorites was the stallion Poplar Byrd, who became Volomite's most successful son in the stud. In 1954 the Larkins bred their greatest horse, the champion Bye Bye Byrd, son of Poplar Byrd and the Billy Direct mare Evalina Hanover. Bye Bye Byrd was trained and raced by Clint Hodgins for the Larkins and was a sensation on the racecourse until his retirement to stud at Poplar Hill.

Upon the death of the Larkins, a dispersal sale was held at Tattersalls, with breeders throughout North America in attendance. Bye Bye Byrd, however, was not included.

Max Hempt, of Mechanicsburg, Pennsylvania, purchased the young stallion in a sealed-bid auction for an undisclosed sum (reportedly half a million dollars). Bye Bye's first crop of foals had raced that season and the Kentucky breeders were not impressed with his initial results—much to the good fortune of Mr. Hempt. In 1973 and 1974 Bye Bye Byrd ranked as the leading sire of money winners in the sport, and through 1976 had sired sixty-eight two-minute performers.

Poplar Hill Farm was also sold and is no longer used as a trotting horse farm.

CASTLETON. The Breckinridge family came over the mountains from Virginia and settled in Kentucky in 1793. Here John Breckinridge, who was later appointed attorney general of the United States by Thomas Jefferson, purchased property and called it Cabell's Dale in honor of his wife, Mary Cabell. At his new home were born his four sons, Joseph Cabell, John, William L., and Robert J., three of whom became Presbyterian ministers. Robert is often credited with keeping Kentucky in the Union, while his brilliant nephew, John C. Breckinridge, was with the Confederacy. The Breckinridges were horsemen as well. William, a Centre College president, once stood the pacer Tom Hal for service and John C. helped to form the Kentucky Trotting Horse Breeders' Association.

A daughter named Mary Ann was given a part of Cabell's Dale when she married David Castleman, in 1812; this farm was later named Castleton. Mary Ann died in 1816 and Castleman later married Virginia Harrison. About 1840 he built the Castleton mansion, where John B. Castleman was born. A Confederate officer and a general in the Spanish-American war, John B. Castleman became the first president of the American Saddle Horse Breeders' Association.

After David Castleman died in 1852 Castleton was owned by several families, including the Higginses, Inskips, and Fosters. The next owner to be a horseman was Stewart W. Ford of Richmond, Virginia. His son, B. W. Ford, who took up residence there about 1885, made Castleton into a modern stud farm but he did not continue for long. The panic of 1893 caused horse values to drop, so Ford went back to Virginia and leased the place to James R. Keene of New York. Keene bought Castleton in 1903. Under his ownership Castleton became the most celebrated Thoroughbred farm in America. Major Foxhall A. Daingerfield, a brother-in-law, was made resident manager. He had a love for trotters, having bred them in his home state of Virginia. One of his products was the fastest trotting stallion in the years following the Civil War. Daingerfield was often seen at the Trots.

Keene himself seldom visited the farm. He just provided the horses that Daingerfield wanted. With outstanding sires and great matrons he soon made the Keene stable of runners the most successful in America, and a stable sent to England also won many victories.

Keene was noted as one of the most spectacular plungers in the history of Wall Street before he had his wings clipped in a losing venture. Then, in 1911, New York had an antiracing crusade and Governor Hughes closed every track in the state. Keene announced that he was not satisfied with the status of racing in this country and had decided to close out his racing and breeding interests. He died shortly afterward without carrying out his intention.

In November, 1911, David M. Look, of New York, visited Castleton and spent a day looking over the establishment with the manager. Daingerfield set a price that Look accepted and the Keene horses were to be removed by the first of the year. The farm of 1,070 acres was now assessed at $119,795. So passed a great Kentucky nursery of the Thoroughbreds and the harness horse industry was the great gainer.

In spite of the fact that owner Keene had been a multimillionaire, Castleton was badly run down. Daingerfield was one of those easy-going southern gentlemen who paid little attention to anything that was not necessary to his breeding operations. When Look took over the farm he found that it would be necessary to spend as much on repairs and improvements as he did for the farm itself. He spent $21,000 on tree surgery alone and also erected the stone fences that now stretch along Iron Works Pike.

At the dispersal of Ardmaer Farm at Raritan, New Jersey, Look bought Bingen and many of the best mares that had been assembled there by William Bradley. He also leased The Harvester for a year as a companion stud for Bingen. Ill fortune dogged Look, for Bingen died after two seasons and later another stallion, Lee Axworthy, died after a single season. The great trotters were buried with the great runners in the Castleton graveyard.

David M. Look inherited his horsemanship from his father, Samuel J. Look, who was among the leading trotting horsemen of the Bluegrass. He was the senior member of Look and Smith, of Louisville, the most prominent firm in the state in selling trotters at private sale. The son began his career with a single mare, Morning Bells. At his father's suggestion, he bred her to a son of Bingen and got Emily Ellen, "the queen of Castleton." The foals and descendants of this mare started Castleton upward in the trotting world. From a mare that had been a present from a friend of Look's, a son of Emily Ellen sired Spencer, winner of the 1928 Hambletonian and Kentucky Futurity and later to become a two-minute sire. At the suggestion of another friend, C. W. Phellis, a daughter of Spencer was sent to Walnut Hall to be bred to Scotland; this was the start of another great speed cross that originated in Kentucky. Phellis felt obligated to buy the first foal from the Spencer-Scotland cross and he got a champion trotting stallion, Spencer Scott, winner of the 1940 Hambletonian.

Because of the run-down condition of the mansion, Look built a clubhouse for his guests, but it burned just after it was completed. Then came the decision to rebuild the old mansion, on which little work had been done for some decades. Some Kentuckians thought the old house should be preserved as it was and merely repaired and made livable. But Look made the new house modern and much larger. For the most part, Look put up his guests in Lexington and the rebuilt clubhouse was used for the employees. During the years that a son, Samuel M. Look, was the Castleton manager he occupied the farm mansion.

Keene had sold when Thoroughbred racing was at a low

ebb. Later, when conditions had improved, there were those who desired to get the refurbished Castleton and return it to its former Thoroughbred glory. After several offers were turned down, the running-horse men finally enlisted a trotting-horse man to make an appointment with Look. The offer was one million dollars, but Look answered, "Tell your patrons that Castleton is not for sale."

For over thirty years Castleton remained one of the glories of the harness horse world. When David M. Look died early in 1945, at age eighty-two, there was much concern about the disposal of Castleton, the birthplace of champions. A little over a month after Look's death, the farm was purchased by the late Frances Dodge Johnson, later to become Mrs. Van Lennep. She was a leading exhibitor at the great horse shows and an expert at both riding and driving. She had ridden Greyhound to the fastest record under saddle and had also driven some of her formidable trotting stable in their preliminary work. She had the nucleus of a good stock farm in her racing stable, so she commissioned her trainer, S. F. (Septer Faith) Palin to start adding mares as fast as the war conditions of the time would allow. Sep was an ideal man for this task, as he not only knew horses but also knew farm management.

On Palin's recommendation one of his reliable assistant trainers, Francis McKinzie, was named farm manager. McKinzie had grown up with Standardbreds on the Senator Farm near Indianapolis, and his ability did much to make Castleton a leader in the production of high-class harness horses. Now retired from farm management, McKinzie lives in Lexington and is regularly consulted by prominent owners for advice in the selection of racing and breeding prospects, and also for his advice on breeding. McKinzie was followed as farm manager by Bill Brown, and he by the present manager, Carter Duer.

Mrs. Van Lennep was an astute horsewoman who knew bloodlines. Only the best stallions were brought to the beautiful new stud barn she had erected. The band of matrons was continually improved and at one time she purchased

Tattersalls sales arena

more than a half million dollars' worth of good mares and foals from Henry Knight. It was not long before the Castleton products were known throughout the world. Not only trotters and pacers but also the best in the saddle horse breed came from there.

Unfortunately, Frances Van Lennep did not live to see the ultimate result of her efforts. By 1973 the yearlings from Castleton were in such demand that they brought the highest average sales price on record to that date and have continued to command record prices since. The farm now exceeds 2,400 acres and the stallion barn has housed some of the sport's most famous sires—Good Time, Florican, Victory Song, Scottish Pence, Ensign Hanover, Worthy Boy, and Bret Hanover. Under the leadership of Frederick Van Lennep, Castleton also continues to race with success on the Grand Circuit.

ALMAHURST. Armed with land grants from Patrick Henry and Benjamin Harrison, James Knight came to the Bluegrass when it was still a part of Virginia. In the early 1780s he built his log cabin in present Jessamine County. The original humble cabin was discovered recently inside an old house that was being dismantled on Almahurst Farm.

The family raised some horses on a small scale, like many Kentucky dirt farmers who only raised a couple for their own racing pleasure or for sale. Several great Thoroughbreds were born there, including Exterminator, lovingly called "Old Bones" by the racegoers. Some great trotters, including Greyhound, Nervolo Belle, Peter Volo, and Volga also came from the old Knight farm.

F. D. "Dixie" Knight bought part of what is now Almahurst in 1870 from the Huggins heirs. He made his home in a building erected about 1800 by James Williams, which he had operated as a tavern. Here the teams were changed on the stagecoach that ran between Harrodsburg and Lexington. This stop for lodgings and refreshments was later called Huggins Tavern. Tragedy struck in 1932 as robbers killed Dixie Knight in an attempt to ransack the old place.

Up to this point the breeding of horses on the old farm had

been on a small scale. When the place was inherited by Henry Knight he named it Almahurst in honor of his wife. From that point on this has been a farm associated with big deals and the production of great horses, both Thoroughbred and Standardbred. Among the first foals sold by Almahurst was Greyhound, a world champion acclaimed as "the trotter of the century." Algernon Daingerfield, for many years secretary of the Jockey Club of New York, once stunned his associates during the poll to select the horse of the year (most of them thought it would be the Thoroughbred Seabiscuit) by voting for the trotter Greyhound. He said that in his opinion Greyhound was "by far the greatest horse before the public."

The first of the big deals came when Henry Knight bought the entire estate of W. H. L. McCourtie. He now had a sire but he had to stand Mr. McElwyn at the trotting track because there was no stud barn at Almahurst. Knight asked an architect to design a stud barn with a porch and large white pillars. He was told that this just couldn't be done. A little later he requested plans for a house of certain dimensions. There were to be four rooms in back and a wide hallway. On the front there were to be two large rooms, to which a porch and pillars were to be added. The architect immediately drew up the plans. After looking them over Knight announced, "Now you have the barn just as I wanted it." It is now the office and stud barn at Almahurst.

Knight was a man who seemed to have an impulse to buy and sell. In 1936 he decided to sell his trotters and go with the runners, but though he did sell most of his trotters he always had a few around the place. The foals from his new venture were outstanding horses. In 1944 he bought 160 horses from Valdina Farm and sold them within three days. He bought the entire breeding stock of Milky Way Farm and one year later had too many horses for Almahurst, so he leased the Ingleside Farm and barn adjacent to the trotting track. In 1947 he acquired the northern part of Coldstream Stud. This had once been a noted trotting nursery and one of its products, the American-bred Muscletone, became the champion of Europe.

The Almahurst stud barn

All this time Knight had been buying trotting stock. In 1950 he had another dispersal and sold his Standardbred holdings to Castleton for half a million dollars. There was a million-dollar deal in 1951 for the 184 horses of the William G. Helis estate. Following this, Knight sold the northern part of Coldstream and paid two million dollars for the southern part of that farm and its horses. He had a silent partner in this deal and he called this part of Coldstream Almahurst Number Two.

In 1956, while he was at the peak of success as a commercial breeder, Knight tired of this massive horse production. At this time he had over 3,500 acres in farm holdings and Almahurst itself had grown to 2,100 acres. Green Hill Farm, where the Knights were living at that time, was sold. They had given up their Chicago apartment in 1941 and moved there. The former Coldstream Stud was sold to the University of Kentucky for $1,123,500 to be used for an agricultural experiment station. A university conference center, known as Carnahan House, is also operated there. In this sale 750 acres came from Henry Knight and 327 acres from Hugh Grant, a prominent Standardbred breeder. When asked about Almahurst, Knight said, "There isn't enough money to buy Almahurst."

Early in 1959 Knight suffered a stroke and died at age sixty-nine. There was a complete dispersal of the horses. Early that fall Almahurst itself was offered for sale in four sections, of which one was sold. At that time the farm totaled 1,300 acres.

During the Red Mile meeting in 1960 it was announced that James Camp had purchased the old farm. His father, Saul Camp, had built up a great racing stable as well as one of the most important Standardbred stock farms in California. In the shipment from the West Coast there were 134 horses. They went by express and required seven big horse pullmans; it was the largest single shipment of horses in history and cost $25,000. Extensive improvements were made at the farm, but though the Camp racing stable did well the average price of the yearlings was not so high. In the early fall of 1963 it was announced that Almahurst had again changed owners.

P. J. Baugh was well known in business, sports, and politics. A North Carolina state senator and textile industrialist, he had a racing stable that earned its own way. The success of the Baugh racing stable had been noted by Camp and he suggested a merger of the two stables. The deal hung fire until Camp called from California and requested a conference in Lexington. This time the proposition was that Baugh buy Almahurst—lock, stock, and barrel. Even the racing stable was included. The talk was in a private room over a meal that had been sent up by room service. The terms were a large down payment and a similar one later. The only writing material at hand was a napkin and the agreement was written on this. Jack Baugh still has that napkin, signed by both parties. The selling price of the horses that were to be weeded out was to set the price for those that were to be retained. The price was never made known but it is probably the biggest deal ever made in harness racing. In order to get this first big draft of horses, the Tattersalls Sales Company moved its winter sale back a full month. In less than three months Almahurst Farm had disposed of over a million dollars' worth of horses through the Tattersalls sales ring.

Baugh spent another fortune on such improvements as a spacious new home for his family and another for the farm manager and his family, three large lakes, miles of blacktop road throughout the farm, and the planting of thousands of trees. The famous old acreage, which has always had at least one trotter on it since 1892, has so increased in reputation that the sales averages rise year after year. Champions and great horses are produced continually. Noted stallions have been in service on the farm for many years, including Mr. McElwyn, Guy Abbey, Blaze Hanover (later transferred to Castleton of Florida), and Little Brown Jug winner Shadow Wave, who died relatively young. Scott Frost, who won the Trotting Triple Crown (Yonkers Futurity, Hambletonian, and Kentucky Futurity) in 1955 stood at Almahurst. Although he became sterile after only two years in the stud, he was the sire of the 1962 Kentucky Futurity winner, Safe Mission. The 1969 Trotting Triple Crown winner, Lindy's Pride, also began his career at Almahurst.

In 1972 P. J. Baugh and his manager, Albert Adams, were called to Toronto, where at the meeting of the Ontario Harness Horsemen's Association an award was presented to Almahurst Farm as the leading breeder of stakes winners in the Golden Horseshoe Circuit, a group of Toronto-area harness tracks. This was the first time that a breeding farm from the United States had earned this honor from the Canadians.

Almahurst is also the first harness horse nursery to be honored by the Commonwealth of Kentucky with a historical marker. It continues to carry the name of Kentucky to world-wide fame in harness racing.

The farms discussed earlier in this book all have long histories as harness horse nurseries. One of Kentucky's newest Standardbred establishments is STONER CREEK STUD, located in Bourbon County, at the northern edge of Paris. For years a prominent Thoroughbred farm, Stoner Creek was purchased in 1964 by Norman Woolworth of New Canaan, Connecticut, and David Johnston of Charlotte, North Carolina, both of whom had long been active in the sport as breeders and drivers. With the wise counsel of their late farm manager Charlie Kenney, the Paris nursery has already made the world record book. One of their 1971 yearlings, the Meadow Skipper colt Good Humor Man, sold for $210,000, the highest price ever paid for a harness horse yearling. The thirty-one Stoner Creek yearlings also brought the highest average price ($26,254) of any Standardbred farm in history at the 1976 Tattersalls Sale.

The sensational trotter Super Bowl is a Stoner Creek product, bred by the farm and sold as a yearling in the 1970 Tattersalls Sale for $20,000. He was a world champion at both two and three, and winner of the Triple Crown in 1972. Super Bowl stands at Hanover Shoe Farm and his first foals raced with much success in 1976.

Stoner Creek is also the home of the fastest trotting stallion of all time, Nevele Pride, with eighteen miles in two

SUPER BOWL (Stanley Dancer)

minutes or faster including his 1:54 4/5 effort at four in Indianapolis. Nevele Pride won more money from racing ($873,328) than any other trotting stallion. From his first crop Nevele Pride, a son of Star's Pride who had a nasty temper to match his blazing speed, sired the two-year-old champion Nevele Diamond. In his second group of foals was Hambletonian winner Bonefish, who banked $309,375. From only his third crop of foals to race, the premier Stoner Creek trotting sire produced his third straight two-year-old trotting king, Nevele Thunder, who banked over $200,000.

Stoner Creek also stands the great pacing stallion Meadow Skipper. The son of Dale Frost was a world or season's champion each of the four seasons he raced for trainer Earle Avery and was an immediate sensation in the stud.* He is the sire of the sport's fastest race colt, Jade Prince, who won the International Stallion Stake over Lexington's Red Mile in 1:54 1/5. He also fathered the fastest filly of all time, Handle with Care; the world champion racehorse Albatross; the wonderful pacing queen Meadow Blue Chip; and the outstanding racehorse Nero.

Since the advent of trotting horses in Kentucky with Ashland, Woodburn, and many other early farms, the Bluegrass has pioneered in futurities and training methods. The great trotting and pacing horses raised here are still drawing buyers from four continents. Without the many years of the Kentucky influence on the breed there probably would not be much in the line of outstanding trotters and pacers today. Orange County, New York, has claimed the title of "the cradle of the trotter," but it was Kentucky that rocked the cradle.

---

* In early 1977 Meadow Skipper became the leading sire of two-minute performers with 100 two-minute sons and daughters. Ed.

9

KENTUCKY'S TROTTING RACES

DURING THE 1840s trotting horse racing had been established in the North but the tracks were all located near metropolitan areas. The only racing south of the Mason-Dixon Line was in Louisiana and the contests there were generally confined to pacers; most of them were Narragansetts or imports from Canada. The first real trotting news was when Lady Suffolk, the "Old Gray Mare" of story and song, was being featured in the *Spirit of the Times*, the first weekly paper in the United States devoted mostly to sports.

In the October 5, 1850, issue of this journal was the initial hint that trotting was about to invade the South. Under the heading of "Correspondence from Lexington, Ky." there was a letter saying: "We have a week's trotting advertised and should there be enough horses to make sport, we may have regular trotting meetings hereafter. Trotting horses have some friends about here, but the racing (Thoroughbred) men, almost to a man, are down on every effort to get up anything in the line of a 'hucking' race, though an impression is given that before many years trotting will be our national sport, if for no other reason than because it is more useful and less costly."

Thus the advent of trotting in the South was announced. Two weeks later the correspondent told of the Fayette County fair at Lexington. There were beautiful ladies, fine horses, fat cattle, and big mules. The highlight of the exposition was a visit by the statesman Henry Clay, and the fairgoers gave him a great ovation. The report also stated that this fair was the first to be held under the auspices of a new society that same summer. The grounds, located about one mile from town, had been lately purchased by the society.

But the report on the first race meeting indicated difficulties. "We have had, too, some eastern sport in the shape of trotting races each afternoon of the week. The purses were gotten up by one or two individuals. The racing was not such as you [northern and eastern sections] have, in point of time, etc., but was very good for out here and I think that this beginning may lead to a great improvement in this branch of sports on the turf." How prophetic was this statement.

The early trotting promoters had problems. There were only four or five horses and each raced every day. A pacing race was offered but was canceled because there were not enough starters. The secretary's report stated: "I will here say that this was an experiment made under the most inauspicious circumstances. The racing men generally being opposed to the meeting we had no place even to give decent exercise until after the running meeting was over, as the trotters were not permitted to enter the course until that time (two weeks from today). I hope, however, that this experience will open their eyes and take off a part of their prejudices. . . . The races throughout have been well attended and the spectators appear well pleased."

One thing of interest occurred on the first day of the races. There were not enough pacers to make a race but there was one pacer on the grounds, Roanoke. He had a slow record and wished to lower it, so there was an event billed as a race against time. It took him three attempts to beat his old record but he was then listed as the winner and "Time" was second. Lexington always has been noted as a place where horses could lower their records and each year at the Trots there are many such trials.

Over the years these trials have carried several names. During the time of the great Dan Patch they were generally called exhibitions. Many of the farms around Lexington reserved a few promising young mares each year to replenish their farm harems. They were not raced, so that their vitality was retained. Such a mare would be trained, but her only official appearance was a joust against time. From this came the term "breeders' records."

At one time many breeders made no attempt to raise horses to race and, instead, their products were trained just for time trial records in order to gain admission to the American Trotting Register. In an attempt to discourage this practice the governing bodies of the sport passed rules requiring that all winners—even those racing against time—should receive trophies. Accordingly, the promoters bought up a large quantity of tin cups, and these were awarded to horses who lowered their records in time trials. This is why it was said that the driver was "tin cupping" his steed when he went alone against time.

The early race meetings at Lexington were hit-and-miss affairs until a group of men got together in 1859 and held a meeting at the old Phoenix Hotel. The result was the founding of the pioneer Lexington Trotting Club. R. A. Alexander, of Woodburn, was elected president and rules and regulations were set up. Women's lib had not yet reached the Bluegrass for one of the rules of the Lexington Trotting Club was that "No female shall be admitted within the course of this Association unless she is under the proper escort of a gentleman."

Shares were sold in the new venture and a mile track was constructed on a site on the present campus of the University of Kentucky. The following August the organization took over the new track. The race meeting given was not only the first for the new club but it also was the first time that trotting races were presented by any legalized organization.

Trotting races won favor and gathered interest until the Civil War. During the war, race meetings were uncertain affairs, for in some cases raiding parties appeared and disrupted the afternoon program, appropriating the contestants for cavalry uses. In 1862 the soldiers even took possession of the track for a campground. The fencing around the track was torn down and used for firewood. After the conflict, conditions were unsettled and many horsemen, as well as spectators, had financial troubles.

Trotting did not flourish during this period and few race meetings were held by the Lexington Trotting Club. It looked dark indeed for the trotters. On February 15, 1873, there was a paragraph in the *Spirit of the Times* that sounded like an obituary. "There is a possibility that the trotting track at Lexington will be sold off in lots, thereby destroying the only public track used by the trotters in that neighborhood. The property belongs to the estate of the late R. A. Alexander and it already has been surveyed and laid off for building purposes. It seems to us that there should be enterprise enough among the trotting men of Lexington and vicinity to put a stop to this by buying the whole tract of land."

The trotting horse people were jarred enough so that a meeting was again held at the Phoenix Hotel in August of that year. The man instrumental in bringing the group together was W. H. Wilson, who had brought George Wilkes to Lexington. Interest was high and over twenty shares were quickly sold at $250 a share. T. J. McGibbon, of Cynthiana, was elected president and before they were finished they had incorporated the Kentucky Trotting Horse Breeders' Association. Many prominent men were among the founders of this new group, including A. J. Alexander and John C. Breckinridge. The latter had been vice-president to James Buchanan and later a major general of the Confederate army as well as secretary of war under Jefferson Davis. Breckinridge never lived to see the great boost that his group had given to the trotters, for he died in 1875, the year of their first race meeting.

The new organization intended to give speedy action and vowed that they "would not be content until a successful trotting meeting had become part of their history." It was too late that year to plan any races and it took a little longer than they expected, but in 1875 they had a track. The Maxwell Springs Fair Association had purchased sixty acres south of the Southern Railway Station and an agreement was made with the Agricultural and Mechanical Fair for the use of their track. This same course is now the site of the Red Mile.

A curved grandstand had, for some unknown reason, been built on the sweep of the first turn. This gave a long straightaway at the finish of the mile, but the horses had to start on the turn. It was almost a decade later that an addition was made farther up the homestretch and the finish line was moved. Along the homestretch of the fairgrounds was a row of cattle barns, where the youngsters discovered an ideal spot to view the races until they were chased away by the track officials.

About the time harness racing started in Lexington the first auction pools appeared. This form of betting has now vanished but it was the most popular in those days. Each horse in a race would be "auctioned" to the highest bidder and the high bidder on the winning horse would receive all the money in the pool. Sometimes the pools would go as high as $50,000. Drivers and owners often bid on their own horses, and when they won the pool might be more than the purse in the race. Because racing in those days was always two or more heats, pools were sold for the entire race, not just individual heats. The auction pool sellers operated in the lobby of the old Phoenix Hotel, sometimes the night before the race, and sometimes the afternoon of the racing program.

One of the big years for the state of Kentucky was 1875. Over at Louisville a horse called Aristides won the first Kentucky Derby at Churchill Downs. On September 28 a new trotting venture opened at Lexington; the first race on that day was a two-year-old trot in which five youngsters answered the starter's call. B. N. Neale of Danville won it with a colt called Oddfellow. This same two-year-old event is now the world's oldest trotting race. It was appropriately named the Lexington and, although it was dropped for a couple of years, it was revived and is now an important part of the International Stallion Stake each autumn.

But the new organization had failed to advertise. Many

people did not even know when the races started. There was a lack of enthusiasm and a slim attendance that showed no spirit in wagering, according to the *Lexington Daily Press*. But the fame of the Lexington meeting was spreading among the trotting people themselves, for in 1882 the great colts from California appeared. The famous trainer Charles Marvin brought some Palo Alto champions to Lexington that year and came back the following year. He beat the best of Kentucky both times.

The lack of attendance caused a steady decline in the financial condition of the organization. It appeared that the Kentucky Trotting Horse Breeders' Association was slated to join the old Lexington Trotting Club in oblivion when several prominent horsemen of the section met in 1886 to devise measures to rejuvenate the trotting races. There was a shake-up in the organization and the most important appointment was that of the able Ed A. Tipton as secretary. Tipton knew the value of advertising and he spread the name of the Lexington Trots in every paper that would accept an ad. The understanding was that payment for this advertising practically depended on the success of the trotting meeting. With ideal weather, great racing, and record-breaking crowds, the races were a huge success. All claims were paid off and a handsome balance was left in the treasury. This was the real start of the Lexington Trots.

In 1889 a correspondent wrote, "The whole range of country from Bangor, Maine, to San Francisco, California, and from New Orleans to Montreal, was represented at this meeting." Kentucky earlier had drawn the whole nation to purchase the Bluegrass trotters and now they were again drawing the nation, this time to see the best horses race. Even today this is typical, for in the fall the trotting horse people gather from all over the continent and mingle with visitors from Europe as well as Australia and New Zealand.

Local people had been relied upon to serve as starting judges up to 1891. They were the ones who tried to get fair starts in the races. If the horses were scattered as they approached the start, or if some driver tried to gain a starting advantage over the rest, then the judge would ring a big recall bell and the horses came back to try again. This was known as scoring. When the drivers had their steeds well lined up in a fair start then the word "Go" was given and the race was under way. It took an able person to handle these drivers; at times it might take as much as forty-five minutes to get a field away. The noted starting judge Frank B. Walker of Ohio was engaged for 1891 by the Lexington organization and was a fixture for many years until his retirement.

The Kentucky Association that owned the running track became financially embarrassed in 1891 and was forced to offer its grounds and track for sale. The Kentucky Trotting Horse Breeders' Association wanted a track of its own and negotiated for the purchase but failed to acquire the property. The failure was probably a fortunate development. It is doubtful that the trotting races could have been such a success on that property. The fairgrounds had better railroad and shipping facilities at that time, and both the horses and the spectators relied on the railroad for transportation to and from Lexington.

The idea of owning their own track lingered and in 1893 the group had prospered to such an extent that they purchased half interest in the fairgrounds. They now had some say in the improvements needed. The mile track was moved and the old grandstand was razed. The grandstand had become unsafe and in fact collapsed during a race meeting. Today the curve of the trees near the old Floral Hall is a faint trace of the outline of the old stand. The track was lengthened and widened so that it was seventy feet at its narrowest point; the result was the present racecourse—the Red Mile. At this same time a half-mile track was constructed in the infield, and a tunnel, just south of the present grandstand, for access to the infield track.* The lumber from the first stand was

---

* It has been stated elsewhere that the tunnel was constructed in 1916. We find, however, that its existence in 1893 was reported in *Wallace's Monthly* 19:494. Ed.

reused to build stables. One was the old barn used by Hunter Moody, later moved to make room for a road. Another was the big barn that today can be seen on the first turn. Originally known as the Roy Miller barn, it was later taken over by Fred Egan.

The spectators who came in 1893 found a new grandstand, a double-decked structure that was one of the best ever known in the sport. From every seat a person could see every inch of the track. This magnificent old stand was constructed at a cost of $25,000, a sum that would hardly build a small house today, and it accommodated over 5,000 people.

On the new track that year was initiated a race first called the Stallion Representative stake. Today it is the Kentucky Futurity, the oldest trotting futurity in existence and now a part of the triple crown of trotting.

A depression hit the nation that same year, so by 1896 the fair was in financial distress. The trotting association paid off all debts and assumed full ownership. The Red Mile had become an independent track.

Improvements continued until there were thirty-five stables and accommodations for 500 horses. A row of stables ran from Broadway to the gate and people stopped to see their favorites before entering the grounds. Fire left gaps in the line and the grooms called it "rat row." The harness repairman was near the gate and so was the blacksmith shop of McMair Kerswill. Several barns were grouped nearby for the trainers and three barns were inside the gate. (This whole group was recently removed.) Several structures were in the northwestern corner and one long barn was along the backstretch. The latter in time developed a list like the Tower of Pisa and as a groom expressed it, "pointed toward Joneses." It was removed as a safety measure. The row of barns along the homestretch and the old Calumet barn are still standing.

In addition to the two tracks there was the "boulevard," a roadway 100 feet wide and half a mile long north of the racetrack. This strip near the tobacco warehouses was a throwback to the speedways found in the larger cities and was used for short bursts of speed in training.

The second grandstand at the Lexington track

A disputed racing decision

The city of Lexington at one time frowned upon pool selling, so these operations were moved to the circular building that stands at the first turn. This is the only remaining trace of the old fair; it is the octagonal Floral Hall built in 1880 by architect John McMurtry. Later used as a stable, it is now the "Stable of Memories," a museum for the preservation of harness horse history. At one time Lexington annexed a section that took in part of the track, but the Floral Hall was just outside the new city line, so the pool selling continued there on the second floor.

It was the Trots that started the tradition of Lexington hospitality. The moneyed men rented houses, complete with servants, in the section east of Transylvania College; carriages with horses and drivers were on call. During the two weeks of fall racing there was a constant round of dinners and parties and it was a social event as well as a sporting event.

Spectators went to the races in carriages. One of the stirring sights was Lamon V. Harkness, founder of Walnut Hall, arriving in a tallyho with a four-horse team, bugler, and footmen. The coach, carrying the family and guests, was driven into the center field. After World War II, it was like a ghost from the past when Mrs. Van Lennep brought her champion hackney pony team in from Castleton. She drove her four-horse team, hitched to a coach, up the track in front of the grandstand.

In time the private parties in town became less frequent, but the farms began to open their doors at the racing season. A lavish meal would be served and later the yearlings would be shown to the guests. In recent years the size (and sometimes the bad manners) of the crowds have discouraged this custom, but at some farms visitors are still entertained at a light lunch before the showing of the yearlings.

Tracks wear out and new dirt must be used in a process called resoiling, which means taking out the dead dirt and replacing it with new soil. This was done at the Red Mile in 1916. The following year the smaller oval was generally overhauled, for storms had eroded the inside lines of the track. Later it was made into a cinder course for wet-weather racing. This was not popular, so now it is again a fine half-mile dirt track.

Lexington continued to hold great races each fall, without any connection with a racing circuit. The oldest and biggest of these circuits was formed in 1871—four years before the Trots—and is known as the Grand Circuit. Originally formed to draw the best horses from the metropolitan areas, a group of tracks in the hinterlands offered good purses so that the rest of the nation could see the best of the horses. Using express trains, the horses went from track to track. Sometimes there were fourteen or more carloads of horses. There was an old saying that a Grand Circuit horse would beat you going 100 yards or 100 miles. Lexington dropped its independent status in 1912 and joined the Grand Circuit.

In 1931 the wonderful old double-decked grandstand burned at the end of the meeting. For the first time since 1875 the Trots were not held at the Red Mile, a name given to the Lexington oval by sportswriter Frank G. Menke. (The name caught on and today is official.) In 1932 and 1933 it was necessary to lease the running track for the fall races. The closest that Lexington came to missing a year of racing was in 1932 when continued rains allowed only the Futurity to be raced. The present grandstand was completed by 1934 and trotting was back again at the Red Mile.

Lexington thrived and continued to have outstanding races. It had weathered the panic of 1893 but was not so lucky in the 1929 depression. Racing was in bad shape before the depression hit and had it not been for the efforts of E. Roland Harriman the sport would have withered and died. To perpetuate the breed, he purchased and revived the register, and he bought the yearbook, a documented chronicle of racing. He also formed the Trotting Horse Club of America to sponsor races and help tracks to keep racing going in the hard times of the 1930s.

When the horses went to Lexington in 1942 there were rumors concerning the Kentucky Futurity. Several good colts were ready to start but before entry time the owners and drivers demanded to know the amount of the purse. They were refused an answer, so no horses were entered and no Futurity was raced. The Kentucky Trotting Horse Breeders' Association went into receivership.

A new organization with a similar name was formed and the Lexington Trots Breeders' Association has carried on. One of its first aims was to restore the Kentucky Futurity. It took from 1943 through 1945 to get the race started again and during those lean years a substitute race was offered for the good colts. Only the three-year-old trotting division was revived.

Many dreams have come true over the years. One of the earliest was in the second Futurity in 1894. Ed Ayres of Duckers, in Franklin County, Kentucky, had a farm and raised trotters. He spent most of his time in a wheelchair, for he was paralyzed from the waist down. The panic had hurt him and it appeared that he might lose his farm. He had kept his fine filly Buezetta eligible for the Futurity and she was drawing prospective buyers. On the day before the race he turned down a $20,000 offer, which would have paid off all debts on his farm. Many thought Ayres was unwise to refuse this but he had faith in his filly. The evening before the race he had the caretaker wheel him over to the stall door and the pair stood vigil all night. The next day Buezetta won and the $22,000 purse saved the old homestead.

A few years later Boralma was favored and just before the race Dr. Scott McCoy sold him with the condition that the sale was not to be consummated until the horse was safely delivered at Lexington, for at that time many horses were injured by the railroad. A special train was hired and McCoy rode in the cab with the engineer to see that his colt had no rough handling. The engineer may not have appreciated the back-seat driving, but the horse arrived in good condition, the sale was completed, and Boralma won the 1899 Futurity.

Another Kentucky-bred horse was Siliko, purchased and trained for just one race that year, the Futurity of 1906. The track was muddy and "Knapsack" McCarthy did not get too close to the inner rail at the start. Ben White hustled The Abbe into the opening and when McCarthy saw this he tried to cut White off. It ended in a collision. For some peculiar reason McCarthy was not censured for his foul and instead the judges sent White to the barn for foul driving. It was necessary to get another driver, but Siliko was the winner. Soon after winning the Futurity, Siliko was sold to Europe, where he was the champion trotter 1908–1910, after which he was repurchased by Mr. Madden and returned to Hamburg Place where he sired some good trotters.

One of the traditional things at Lexington had been to have the winning driver sit in a floral wreath after the victory. It was a bruised and bandaged McCarthy who was surrounded by the flowers that day. Knap always took great care in hiring men. Belts and cigarettes were not allowed, for he claimed that the men would spend the whole day either pulling up their pants or rolling those "paper pipes."

Tobacco magnate W. N. Reynolds had a farm that was just behind the grandstand. When the late Gibson White was putting up a valiant battle against tuberculosis "Uncle Will" approached the boy's father, Ben White, with a suggestion that at $200 an acre the farm would be a great thing for the boy. It finally wound up in a partnership between Gibby and Reynolds's secretary, a man named Long. Later the farm was leased to Henry Knight when Almahurst was not large enough to handle his Thoroughbreds, and the twenty-stall barn was included in the lease. Later this barn burned but the house at Inwood Farm remained the home of the Whites and can still be seen in the trees on the far turn. The increasing number of cars at the track created a problem, so the Lexington organization purchased the Ingleside Farm, sold it, and retained a few acres for a parking lot and for the long row of modern stables that now mingle with a few from the early trots.

In 1889 the Transylvania, a race for the fastest trotters, appeared. In 1906, after Nut Boy won, his owner went down to sit in the floral wreath; this was the actress Lotta Crabtree. Another prestige event was and is the Walnut Hall Cup, first raced in 1897. Joan was the 1910 winner, driven by Mike McDevitt, a big Irishman whom Captain David Shaw had

Winter at The Red Mile

Barn built with lumber from the old grandstand

picked out of his steel mill. Mike became an accomplished reinsman; although he had a hand like a ham he had a touch like a feather. Doc Dippy was a character who sold stop-watches and McDevitt could drive him crazy, for Mike was an accomplished pickpocket. When the crowd was looking at Dippy's watches McDevitt would manage it so that before he was done Dippy's watches would be in people's pockets and everybody would be carrying a watch that belonged to someone else.

Another feature that disappeared but has since been restored to the program is the Tennessee Pace. The initial winner was Star Pointer, the first horse to break the two-minute barrier. Another winner was Dan Patch.

One group of Lexington trainers belonged to a profession no longer known. These men just trained horses for time records and had permanent quarters at the track. In the fall they could be seen driving colts at full speed in hopes of attracting a buyer.

There was racing at the Kentucky fairs but the advent of extended night race meetings sounded their death knell. W. H. Wilson, after he lost George Wilkes, bought a place near Cynthiana named Abdallah Park. Fairs and races were given but the people had to travel several extra miles around a big hill to get to the track. To remedy this, Wilson cut a roadway through the hill so that the grounds were just a short trotting stride from town.

Officially, the fastest track in the nation is the Red Mile and the speed shown each fall is almost unbelievable. A pacer went a time trial during Kentucky's first day of racing and a pacer has recorded the fastest mile in history in a time trial at Lexington: Steady Star paced a mile in 1:52 during the 1971 meeting. During the races there is a continual series of these trials and no track can come close to the total of 1,142 two-minute miles that this famous old oval had seen at the end of 1976.

No longer do the trots feature parades of famous stallions from Lexington's celebrated stock farms and no longer are the mares honored, like the champion Nancy Hanks or Elizabeth, whose foals were all brought to Lexington for the occasion. Even Greyhound came out of retirement from Illinois to join the group on the track. No longer is there a parade of the yearlings past the old Lafayette Hotel and the Phoenix, though these expensive youngsters can now be seen in their section at the Tattersalls sales building.

Kentucky now has four extended night parimutuel meetings—at Lexington, Latonia at Florence, Louisville Downs, and Midwest Harness at Henderson. But the real feature is the fall Grand Circuit meeting, for that is when the horsemen gather. From all over the nation they bring their horses, because winning at Lexington carries extra prestige, and there is spirited bidding at the greatest yearling sale in the nation. To sum it up—the harness horsemen still look to Kentucky, just as they did when Woodburn was its leader.

A Note on Sources

As a result of his lifetime association with the sport, Ken McCarr possessed a great deal of unpublished information on the history of the Standardbred horse and that of harness racing. Without doubt, he drew liberally on that knowledge in preparing this book. He also mentioned, as rich sources of information on Bluegrass farms and horses, his friends Robert B. Jewell and the late Jesse Shuff.

The author possessed a very extensive collection of periodicals and books connected with Standardbred history. Such journals as *Wallace's Monthly, The Spirit of the Times, The Horse World, The Chicago Horseman,* and the *Horse Review* chronicled the formative period during which the American harness horse gradually attained the status of a separate breed. Also useful are old files of such extant journals as *The Horseman & Fair World, The Harness Horse,* and *Hoof Beats.*

Among the books that should be consulted by the reader who wishes to pursue this topic are John Hervey's classic *The American Trotter* (New York, 1947), Hamilton Busbey's *Recollections of Men and Horses* (New York, 1907), *The Two-Minute Horse* by James Clark (Cleveland, 1922), and the works of W. H. Gocher.